Beyond the Quantum

Beyond the Quantum

A Quest for the Origin and Hidden Meaning of Quantum Mechanics

Antony Valentini

OXFORD
UNIVERSITY PRESS

Great Clarendon Street, Oxford, OX2 6DP,
United Kingdom

Oxford University Press is a department of the University of Oxford.
It furthers the University's objective of excellence in research, scholarship,
and education by publishing worldwide. Oxford is a registered trade mark of
Oxford University Press in the UK and in certain other countries

Published in the United States of America by Oxford University Press
198 Madison Avenue, New York, NY 10016, United States of America

British Library Cataloguing in Publication Data

Data available

Library of Congress Control Number: 2024952133

ISBN 9780198853749

DOI: 10.1093/oso/9780198853749.001.0001

Printed and bound by
CPI Group (UK) Ltd, Croydon, CR0 4YY

Links to third party websites are provided by Oxford in good faith and
for information only. Oxford disclaims any responsibility for the materials
contained in any third party website referenced in this work.

The manufacturer's authorised representative in the EU for product safety is
Oxford University Press España S.A. of El Parque Empresarial San Fernando de Henares,
Avenida de Castilla, 2 – 28830 Madrid (www.oup.es/en or
product.safety@oup.com). OUP España S.A. also acts as importer into Spain
of products made by the manufacturer.

To my father

Acknowledgements

First and foremost I thank my parents—Maria and Michele Valentini—for their lifelong love, support, and inspiration.

I am also grateful to those mentors, colleagues, and friends who in various ways supported this work—in Italy, England, Canada, France, and the USA—over some three decades. In chronological order: Dennis Sciama, George Ellis, Bruno Bertotti, Marcello Cini, James Cushing, Harvey Brown, Jeremy Butterfield, Simon Saunders, Lee Smolin, Christopher Isham, Howard Burton, Lucien Hardy, James Hartle, Roger Penrose, Carlo Rovelli, Jonathan Halliwell, Philip Pearle, Peter Barnes, and Murray Daw.

It is a pleasure to recall lively discussions with Sebastiano Sonego, during the long Italian summer of 1990, when these ideas were first taking shape.

It is also a pleasure to thank those students and collaborators who helped push these ideas forward in more recent years, in particular: Samuel Colin, Adithya Kandhadai, Patrick Peter, Nicolas Underwood, and Sandro Vitenti.

I am grateful to Philip Pearle for comments on the Prologue, to Chris Dewdney for kindly fact-checking parts of Chapters 2 and 3, to Carlo Rovelli for comments on Chapter 8, and to Lee Smolin for a sympathetic reading of Chapters 8 and 9.

I owe a special thanks to my wife Vivien, and to my stepdaughter Sophie, for their companionship while writing this book.

Finally, I am grateful to Sonke Adlung at Oxford University Press, for his warm encouragement and helpful advice, and to Giulia Lipparini for sensitively shepherding the manuscript towards production.

This book is based on 35 years of thinking and research. The Austrian physicist Erwin Schrödinger was perhaps exaggerating a little, but even so had a point, when he wrote:

> If you cannot—in the long run—tell everyone what you have been doing, your doing has been worthless.

Contents

'But there are *gradations* of matter of which man knows nothing; the grosser impelling the finer, the finer pervading the grosser. . . . The ultimate, or unparticled matter, not only permeates all things but impels all things—and thus *is* all things within itself.'

—Edgar Allan Poe

Prologue

What is real?

'Quantum' has become a byword for the strange and incomprehensible, the paradoxical, and the indisputably weird. It has been said that nobody understands it. And yet quantum mechanics is the most fundamental theory of science (first applied to atoms and molecules). It increasingly powers modern technology—from the laser to the quantum computer. How can our most basic theory of the world be so impenetrable? It is as if we have hit a roadblock in our millennia-long quest to understand nature: in the quantum world, no one knows what reality is like.

The burning question—what is real?—is wide open. At the deepest level of science, no one can answer it with confidence. This should be a cause for excitement and a spur for urgent scientific investigation. But not only has modern physics been unable to establish what quantum reality is like: the very *idea* of 'reality' has been left by the wayside.

In the early twentieth century, when quantum mechanics was being developed, the idea of reality came to be widely derided as naïve and old-fashioned, a relic of the nineteenth century as quaint and outmoded as the steam engine. In our sophisticated age we know (or think we know) that reality is unknowable and pointless to think about. To ask 'what is real?' is to waste time. Persistent attempts to understand quantum reality are regularly dismissed as 'just philosophical' and of no value.

To those who expect science to tell us about the world and how it works, all this may seem especially egregious and concerning. And so it is. But science has been here before. In the words of the Italian Renaissance historian and statesman Francesco Guicciardini:

> Past things shed light on future ones; the world was always of a kind; what is and will be was at some other time; the same things come back, but under different names and colours; not everybody recognises them . . ..

Beyond the Quantum. Antony Valentini, Oxford University Press. © Antony Valentini (2025).
DOI: 10.1093/oso/9780198853749.003.0001

The eternal conflict: perception vs. reality

This is a conflict that echoes down the ages, from ancient times to the present day. There have always been those who emphasise our sensory perceptions—what we can see, hear, and touch—and those who emphasise instead the reality behind the scenes, a reality which we perhaps cannot see, hear, or touch. Support for one position or the other has waxed and waned over the centuries. The conflict arises from a deep division within the human mind. Which side is taken sometimes seems to come down to a question of character or psychological disposition. People tend to be drawn to one side or the other—sometimes passionately so—often without knowing quite why.

Here is a good example. Around the year 1900 a philosophical battle was raging over the reality of atoms and molecules. Bizarre as it may sound today, even though chemists had long ago worked out the detailed compositions and shapes of molecules (such as the famous 'benzene ring'), physicists were still arguing over whether atoms and molecules existed at all.

The focus of the controversy was 'thermodynamics'—the science of heat—whose laws had been well established. Those laws could be applied to calculate things like the change in pressure of a gas in a container. If the gas was heated, the pressure went up. If the container expanded, the pressure went down. All this could be seen in the laboratory and the details fitted the calculations perfectly. But a question remained, which thermodynamics did not even try to answer: what *causes* a gas to exert a pressure in the first place?

For much of the nineteenth century many scientists thought the question pointless. But some dissidents had developed an answer: a gas exerts a pressure on the walls of the container because the gas is made of zillions of tiny invisible molecules moving at high speed and bouncing repeatedly off those walls like tiny tennis balls (Figure 1). Zillions of bounces of zillions of tiny invisible molecules add up to a pressure we can notice and measure. The hotter the gas, the faster the balls move, meaning more impacts and more pressure. So the theory gives an explanation for what we call 'temperature' as well as pressure.

The calculations of this 'kinetic theory of gases' gave the same results as thermodynamics and so agreed just as well with experiment. But was kinetic theory true?

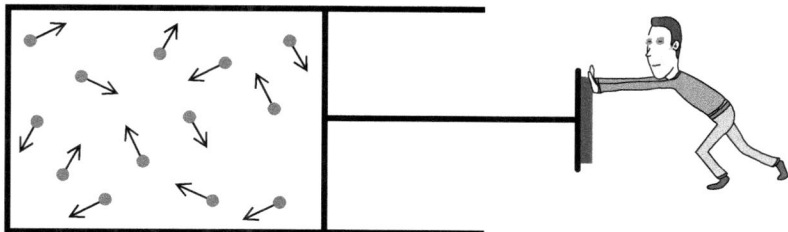

Figure 1. A cylinder of gas exerts a pressure. Where does the pressure come from? The kinetic theory of gases tells us that the gas is made of zillions of tiny invisible molecules bouncing around—and that the pressure is caused by zillions of tiny impacts on the walls of the cylinder.

Some argued that this picture of invisible bouncing balls was so much fantasy and fiction. In those days there were no electron microscopes and no images of atoms or molecules. The reality posited by kinetic theory could not be seen directly. Why play games inventing things we cannot even see? Why not just accept the laws of thermodynamics as they are? After all, unlike kinetic theory which dabbles in the invisible, thermodynamics deals only with visible things we can measure—like the pressure in a big cylinder of gas in the lab. Thermodynamics is proper science; kinetic theory borders on philosophy and metaphysics. The kinetic theorists countered that in science we must look for explanations and not simply describe what we see, and that good explanations often require us to invent things we cannot currently see. In short: the thermodynamics faction said that as scientists we should stick to describing our perceptions (what is visible to our senses), while the kinetic theory faction said that we should also try to describe a deeper (perhaps invisible) reality that can *explain* our perceptions.

Today it is difficult to appreciate or comprehend the passions that were aroused in this historic debate. Even the Russian revolutionary Vladimir Lenin weighed in, vigorously arguing in his book *Materialism and Empirio-criticism* of 1909 that the conflict was an ideological battle between philosophical 'idealism' (everything is mind) and philosophical 'materialism' (everything is matter). At a time when materialists were still likely to face hostile accusations of atheism, these were more than mere academic disputes.

Famously, two distinguished Austrian professors at the University of Vienna—Ernst Mach and Ludwig Boltzmann—were standard bearers for the opposing camps (Figure 2). Mach had spent his life developing an entire philosophy in which physics should only talk about sensory perceptions (or 'sensations'). For Mach science should, for example, understand gases only in terms of things we can directly see and measure. Boltzmann, in contrast, had spent his life developing the kinetic theory, and had something of his own philosophical position in which 'atomism' was the basis of true scientific explanation. Boltzmann was a man with a passion, and with an idea that would change the world. He committed suicide in 1906, just as laboratory evidence for the existence of atoms was beginning to emerge. Mach died in 1916, to the end denying the kinetic theory and disparaging those who believed in invisible atoms and molecules.[1]

Today—when physicists work with materials such as graphene which are only one atom thick, and biologists analyse the detailed molecular structure of living organisms—Mach's opposition to the atomic theory

Figure 2.   Perception versus reality in thermodynamics. Mach (left) argued that invisible atoms were metaphysical fictions, while for Boltzmann (right) atomism was an essential foundation for scientific explanation.

[1] As late as 1910, in response to criticism from the German physicist Max Planck, Mach wrote: 'If belief in the reality of atoms is so important to you . . . I decline with thanks the communion of the faithful.'

seems quaint and even absurd. Mach's influence no doubt impeded the acceptance of Boltzmann's ideas, and may well have contributed to Boltzmann's suicide. Why were so many people so vehemently opposed to invisible atoms as an explanation for what they could visibly see?

Perception vs. reality in astronomy: the Copernican revolution

The recurring conflict between perception and reality can also be found in theories of astronomy. The 'Ptolemaic system' of the ancient Greeks was named after Claudius Ptolemy, who around the year 150 CE published his definitive book—which later came to be known as the *Almagest* ('the greatest'). Ptolemy's system of astronomy was taught at universities and academies throughout Europe for some 1400 years. The earth was at rest at the centre of the universe. The Sun, the Moon, the stars, and the other planets were in orbit around the earth. We know this to be wrong, but in retrospect it seems an understandable mistake. The mistake did not begin to be widely corrected until 1543 when the Polish astronomer Nicolaus Copernicus published his book *On the Revolutions of Heavenly Spheres*. This momentous event triggered the Copernican revolution: humanity came to understand that the earth was *not* at rest at the centre of the universe but was merely one of several planets orbiting the Sun (Figure 3).

More insidiously, however, Ptolemaic astronomy divided the universe into a physical 'earthly' realm and an abstract 'heavenly' realm. It was

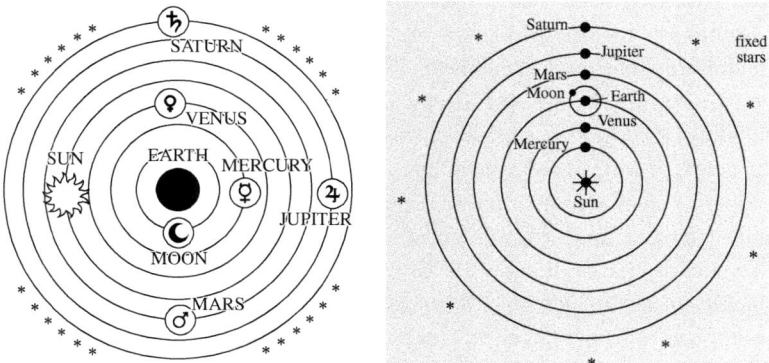

Figure 3. The Ptolemaic system (left) with the earth at the centre of the universe, and the Copernican system (right) with the Sun at the centre of the universe.

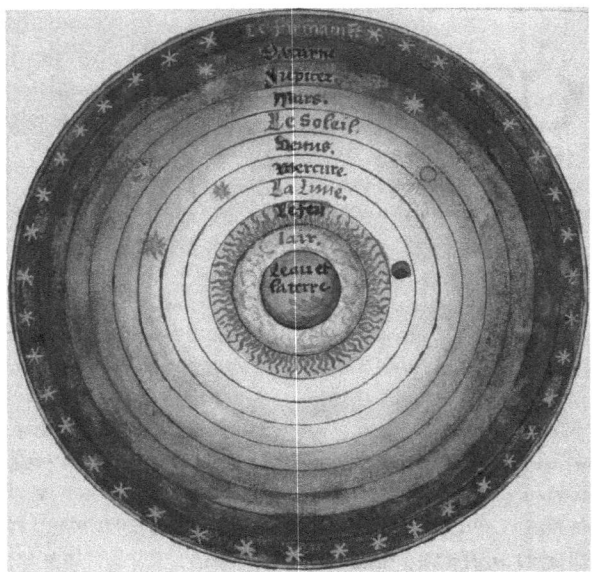

Figure 4. The divided universe of Ptolemaic astronomy. The inner region below the Moon is made of ordinary matter (earth, water, air, and fire) and is subject to change and decay. The outer region above (and including) the Moon is made of heavenly ether and is unchanging and eternal. (From *L'Esphere du Monde* by Oronce Fine, 1549.)

believed that the heavens could not be understood in earthly terms. This belief had been advocated by the Greek philosopher Aristotle in his book *On the Heavens* (350 BCE). The dividing line was the orbit of the Moon. Everything below the Moon was made of ordinary matter—earth, water, air, and fire—obeying earthly laws. This 'sublunary sphere' was imperfect and subject to change and decay. In contrast, everything above (and including) the Moon was made of heavenly 'ether', obeying heavenly laws. The celestial world was perfect, unchanging, and eternal (Figure 4).

Ptolemy also assumed that in a perfect world there could be only one kind of motion: uniform motion in a circle. This belief had, again, been advocated by Aristotle, as well as by the Greek philosopher Plato in his *Timaeus* (circa 360 BCE). For this reason, in Ptolemy's system all heavenly motion had to be explained in terms of 'perfect' uniform circular motion. This had a bizarre consequence. As seen from Earth, a planet in the night sky follows an irregular motion: it moves along a path, stops and backtracks for a bit, and then continues on its way. We now know this is because both the earth and the planet are moving. But Ptolemy believed that the earth was at rest and that the only allowed motion in the heavens was uniform and circular. How, then, to explain the irregular—forward and backward—motion of a planet in the sky? Ptolemy achieved this by *combining* different uniform circular motions: a circular planetary orbit together with an 'epicycle' around that orbit (Figure 5).

But the details of the Ptolemaic system are less important here than the philosophical outlook which underpinned it. The *sole purpose* was to predict the observed motions of heavenly bodies across the sky—the stars and planets, the Sun and the Moon. Those who had mastered Ptolemy's mathematical system could, for example, predict the date and time of the next solar eclipse with uncanny accuracy. It was, however, not done to ask physical questions about what caused the planets and other bodies to move. Nor did it matter if the epicycles were regarded as physically real or as mere mathematical fictions. Never mind what was *really* happening in the sky! It

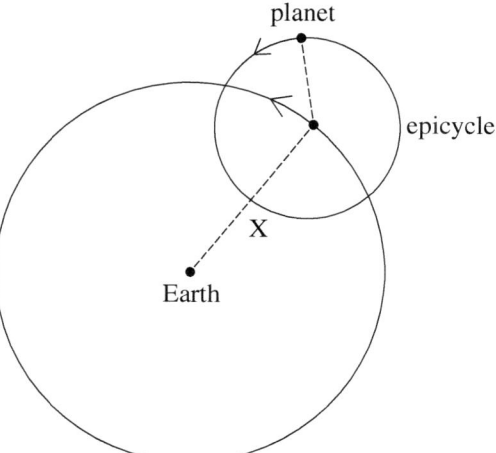

Figure 5. Simplified example of an epicycle. In Ptolemaic astronomy the earth is at rest at the centre of the universe. The motion of a planet is a combination of *two* (uniform) circular motions. As seen from Earth, as the planet approaches the point X it will appear to slow down and will temporarily move backwards across the sky. (In Ptolemy's system the circular planetary orbit is not centred exactly at Earth as shown here but is slightly offset, to account for the planet moving further from and closer to Earth as it travels.)

was after all believed that the heavens could not be understood in the same physical way as earthly things, and therefore all we could do was invent mathematical rules that correctly predicted what we saw. For as long as it worked, why ask questions?

In ancient times, this philosophical position was commonly referred to as 'saving the phenomena'. In plain English this meant that as long as we are able to predict what we see in the sky, we have succeeded—regardless of whether we understand what is really happening. This was a mirror image of the attitude that became fashionable in theoretical physics—some two thousand years later—in the early twentieth century. Concern yourself only with sensory impressions and with making successful predictions. Do not waste time asking questions about what is really happening. Workers in theoretical physics today sometimes sum up this creed as 'shut up and calculate'. This pithy slogan may have a modern ring to it, but a similar attitude once dominated the teaching of Ptolemaic astronomy.[2]

It may seem incredible to us now, but for centuries the most sophisticated and technically accomplished scientists, philosophers, and mathematicians believed that the stars and planets inhabited an abstract and unknowable realm beyond earthly comprehension. Who were we—mere mortal and limited earthly creatures—to ask why of the eternal heavens?

It took the genius and daring of the German astronomer Johannes Kepler—in the early seventeenth century—to start thinking deeply about the orbiting planets as if they were physical bodies like the earth. Kepler was a convert to the Copernican system. In his *New Astronomy* of 1609 he reasoned that, if the planets do indeed orbit the Sun, then presumably the Sun itself is the source of some power or influence—a 'motive force'—that makes the planets move. Kepler argued that this force would be weaker for a planet that is further away from the Sun, making it move more slowly—and stronger for a planet that is closer to the Sun, make it move more quickly. Kepler's mathematical understanding of forces and motion was in some respects mistaken, but even so his physical thinking about 'the heavens' paved the way for modern science and astronomy.

In his monumental *Principia* of 1687, the great English scientist Sir Isaac Newton gave an accurate account of what we now call 'gravitational

---

[2] As famously highlighted by the distinguished French physicist, historian, and philosopher of science Pierre Duhem.

force', which indeed weakens with distance (as the inverse square). Newton understood that forces cause acceleration (change of velocity), and he codified his understanding in terms of three laws of motion. On this basis Newton showed how the Sun's gravitational attraction could explain the elliptical—not quite circular—orbits of the planets which Kepler had described (Figure 6).

In three giant leaps—taken by Copernicus, Kepler, and Newton—humanity emerged from the narrow confines of Ptolemy's sublunary sphere and into the vast universe beyond (Figure 7).

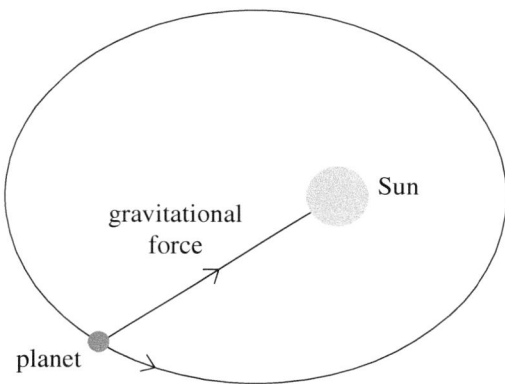

Figure 6.   A planet orbiting the Sun. According to Kepler, and as developed by Newton, the Sun exerts a gravitational force on the planet. The orbit is not circular but elliptical.

Figure 7.   Perception versus reality in astronomy. From left to right: Ptolemy, Copernicus, Kepler, and Newton.

Perception vs. reality in quantum physics: undoing the Copernican revolution

History, as they say, repeats itself. In the 1920s the conflict between perception and reality reared its head once more—this time in quantum physics. On the side of perception we find the Danish physicist Niels Bohr, the German physicist Werner Heisenberg, and many others. On the side of reality we find the great German physicist Albert Einstein, the Austrian physicist Erwin Schrödinger, and very few others (in particular the French physicist Louis de Broglie, of whom more later). The history books tell us that the correct interpretation of quantum mechanics was finally settled by Bohr and Heisenberg, in the teeth of rearguard action by a few diehards like Einstein, who clung to outdated ideas about reality (Figure 8).

In fact, the interpretation of quantum mechanics was *not* settled. And physics today is still locked in the same conflict. Many physicists claim that quantum mechanics is and should be only about what we can see and measure (our perceptions). 'Realists' claim instead that to understand quantum mechanics we need to know what is actually happening behind the scenes—even if the details are currently invisible to us. Like Boltzmann before them, quantum realists often find themselves accused of peddling metaphysical fictions.

The problem begins in the textbooks. University physics students are taught to see the world in a way that is eerily reminiscent of the Ptolemaic system, with a division between a sublunary or earthly realm which is knowable and a heavenly realm which is not. In modern physics the knowable realm is the everyday world of human experimenters with their

Figure 8. Perception versus reality in quantum physics. From left to right: Bohr, Heisenberg, Einstein, and Schrödinger.

macroscopic laboratory equipment, while the unknowable realm is the microscopic 'quantum system'.

Think of an experimenter trying to probe inside a box containing an atom or a single molecule. The laboratory can be described clearly: here, for example, is a Geiger counter (a device that clicks when it detects a radioactive decay), there is a computer screen connected by wires to the Geiger counter, here is the experimenter pressing the on-off switch, and so on (Figure 9). If the experimenter hears a click from the counter and sees a record of the click displayed on the computer screen, no one argues about whether all this really happened. The evidence of our senses is there for all to see and hear. We can say that the laboratory occupies the everyday macroscopic world of *definite states*: either the counter clicks or it does not, either a data point has been recorded or it has not, either an experimenter sees something or they do not. In dramatic contrast, the box containing the microscopic atom or molecule seems to occupy another kind of world altogether. If we measure it with a Geiger counter, a radioactive atom will indeed either decay or not decay. But if left alone and unmeasured the atom can be in an *indefinite state*, according to which it has both decayed and not decayed—a fuzzy 'superposition' of two different states of reality. Or at least: so say the quantum textbooks.

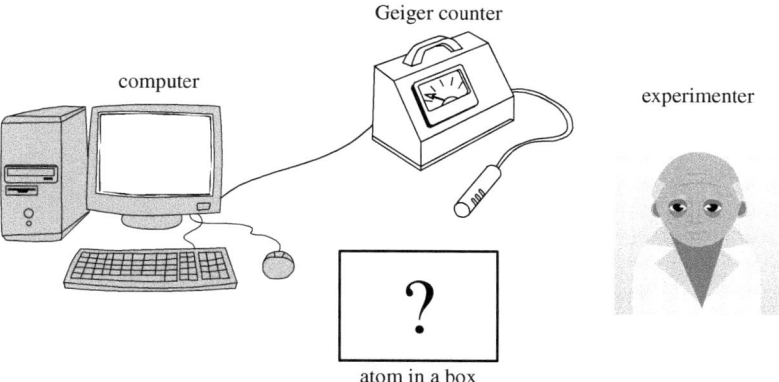

Figure 9. The divided universe of quantum mechanics. The laboratory containing the experimenter and the equipment exists in a definite state. The box containing the unmeasured atom (or molecule) can exist in an indefinite state—an ethereal state of fuzziness beyond human comprehension.

The fuzziness can infect larger systems too. Famously, in 1935, Schrödinger imagined a cat that is poisoned or not poisoned, depending on whether a radioactive atom has decayed or not decayed. If no one is around to observe it, according to quantum mechanics, the cat will then also inhabit the fuzzy realm of indefinite states. Schrödinger's cat can be *both dead and alive* (Figure 10).

The boundary between the two kinds of worlds can be shifted, but it is always there. On this side, where we talk about what we have seen and measured, we can say clearly what happened: the atom has decayed (if we measured it), the cat is dead (if we looked at it), and so on. But on the other side, for unobserved and unmeasured quantum systems, there is a fuzzy and surreal world of which we cannot speak clearly—and about which we cannot even *think* clearly.

Students are taught not to worry about this. After all, quantum mechanics makes accurate predictions. If we want to know the 'half-life' of a radioactive atom—roughly speaking how long it takes, on average, for the atom to decay—then we can apply quantum mechanics to work it out. But do not expect the theory to tell us exactly when a particular atom will decay—or why. The theory only gives us probabilities, or statistics, like the weather forecaster telling us there is a 25% chance of rain tomorrow. It is even worse than that: the weather forecaster at least gives us an explanation in terms of warm and cold fronts, regions of high and low pressure, which are up there in the atmosphere but which are too complicated to model and predict with perfect accuracy. The quantum forecaster does no such thing: if a student asks what is *really happening* when an atom decays, they will be

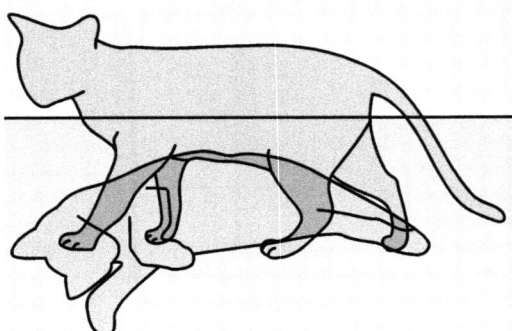

Figure 10. Schrödinger's cat. According to quantum mechanics an unobserved cat can be both dead and alive.

told that such questions are 'meaningless'. All we can talk about with any clarity is what we see when we make a measurement—whether the Geiger counter clicks, or not. Just shut up and calculate the chances of us seeing a click. If the Geiger counter is switched off and the experimenter is asleep at home, there is nothing concrete we can say about the lonely atom in the laboratory. Do not even try to think about what the atom might be really doing when no one is watching—or, for that matter, what Schrödinger's cat might be really doing when no one is watching.

According to quantum mechanics, an atom left alone by itself generally has an indefinite state of reality. The atom acquires a definite state of reality only when it is observed or measured. In the usual jargon: when we measure an atom, it 'collapses' from an indefinite state to a definite one. It is as if a well-defined atomic reality comes into being only when we observe it.

Now all this applies not only to atoms but also to cats and other large objects—in principle, to the whole world around us. We then reach a startling conclusion: *without an observer there is no definite reality*. Or at least: so the theory seems to imply.

Quantum mechanics effectively undoes the Copernican revolution, in two profound ways. First, it restores the ancient division between knowable and unknowable realms: today, atoms are as unknowable as the heavens were in ancient times. Second, human beings are in effect again at the centre of the universe: the very existence of the world, with a definite state of reality, depends on the presence of human observers. If we were not here to see it, the vast universe opened up by Copernicus, Kepler, and Newton could not be meaningfully said to exist.

Like ancient astronomers talking about the heavens, modern scientists often talk about the quantum world as if it were more mystical than physical—as if reality somehow depended on the human mind or on human consciousness. Something, somewhere, has gone terribly wrong.

Why reality won the day

Today we know that Boltzmann was right and Mach was wrong: atoms and molecules really exist, and we have learned more about them than Boltzmann could ever have dreamed. Today we also know that Ptolemy was wrong: the heavens are not unchanging and eternal, they are made of the same kind of stuff we see here on Earth and they obey the same physical laws. In these two historic conflicts, reality won and perception lost. Why?

In part, it is human nature to be inquisitive. Who can observe an explosive chemical reaction between two gases without wondering what is really happening in there? Or gaze at the night sky without wondering what is really up there? This sense of wonder and mystery has driven the scientific quest since time immemorial. People will keep asking questions—unless they are taught not to. And even when they are taught not to, there will always be those who refuse to toe the line.

Another reason has to do with reality itself. Anyone with any experience of life knows that lies have short legs and the truth will come out eventually. In the same sense, when mistakes are made, reality has a way of getting through in the end—however much people may try to deny it. In science, however, this process can take a long time.

In the case of thermodynamics, the reality is that the dividing line between the visible and the invisible is imprecise and ultimately bogus. Imagine a grain of sand so small that it cannot be seen with the naked eye and yet it can be seen with a simple optical microscope. Is it visible or invisible? And more importantly: is it a visible object subject to the laws of thermodynamics—or is it an invisible object subject to the laws of kinetic theory? This is a good question.

Such a situation arose historically when scientists looked at pollen grains suspended in water. Pollen grains cannot be seen with the naked eye, but they are visible under a microscope. We can watch the grains burst, releasing tiny granules. If one of the granules is moving in liquid water, then according to thermodynamics it should be slowed down by friction and eventually come to rest—just as a motorboat with the engine switched off eventually comes to rest on a steady lake. But that is not what happens. Under the microscope we see that the granules are forever shaking and jittering around (even if the water appears perfectly still). They never stop moving. This is 'Brownian motion' (Figure 11, left), named after the Scottish botanist Robert Brown who observed it in 1827. Brown wondered if the granules might be alive. But when he saw the same jittering in particles of very fine dust he was at a loss for how to explain it. We now know that the tiny granules keep jittering because they are constantly bombarded by invisible water molecules (Figure 11, right). Sometimes more molecules are bouncing off one side, sometimes more are bouncing off another side, causing each granule to zigzag around seemingly at random. The details were worked out by Einstein in 1905, providing the first direct physical (as opposed to chemical) evidence for the existence of atoms and molecules.

The case of Ptolemaic astronomy is also instructive. The assumed dividing line between the earthly and heavenly realms also turned out to be bogus. Imagine sending a rocket out into space. For as long as it is below the orbit of the Moon, it remains a real everyday rocket made of ordinary changing matter—and piloted by a mortal and ageing astronaut. What happens when it reaches the lunar orbit? Does it stop in its tracks? Or suddenly turn into ether and partake of the eternal and unchanging? And how sharp is the boundary—an inch thick, a mile wide, or of no thickness at all? We now know that, in reality, there is no boundary. The laws of physics are the same both above and below the Moon.

A star, for example, is not an eternal ethereal being as some ancients believed but a nuclear-powered fire that happens to burn for a very long time. Like fires on Earth, stars can flare up or go out. In the year 1572, astronomers saw a 'new star' appear in the sky. It shone brightly in the heavens, in a small patch of sky where before there had been nothing to see. We would now call it a supernova (or exploding star). The new star was a message, telling all who would listen that the heavens were not immutable but also subject to change like the earth.

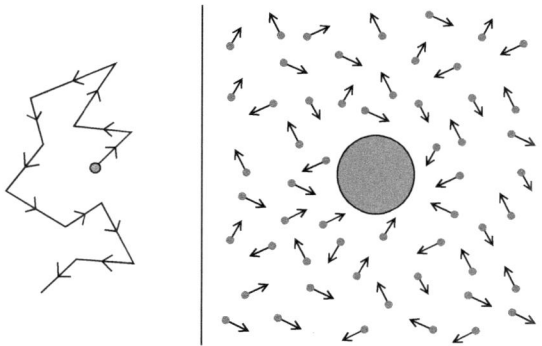

Figure 11. Where the visible meets the invisible. As seen under a microscope, a tiny granule zigzags or jitters around seemingly at random (left). In 1905 Einstein showed how this could be explained by the bombardment of invisible water molecules (right).

Figure 12. Sketches of the surface of the Moon (left) published in 1610 by Galileo (right).

There was more. In 1609 the great Italian scientist Galileo Galilei pointed a telescope at the night sky. He saw that the Moon had mountains (Figure 12). Contrary to what the ancients had believed, the Moon was not a perfect ethereal sphere but an imperfect lump of rock. He also spotted four moons orbiting the planet Jupiter, like a miniature solar system, showing that not everything revolved around the earth. The following year Galileo published these and other findings in his startling and dramatic little book *Sidereus Nuncius*—perhaps best translated as *The Message from the Stars*—which dealt severe body blows to the Ptolemaic system.

Under the weight of such evidence, the idea gained ground that 'heavenly bodies' could be understood as physical systems not unlike those found on Earth. Once this step had been taken, the imagination and insight of Kepler and Newton could do their work. After a slumber of some one and a half thousand years, humanity awoke to the vast reality outside the earth.

Why reality will win again

The sceptics who would put ultimate limits on human knowledge and understanding were defeated in the aftermath of 1543 and again in 1905. There is every reason to expect they will be defeated once more.

As in previous centuries, the sceptics of quantum physics try to draw a dividing line between the knowable and the unknowable. And once again the dividing line is imprecise and ultimately bogus. On one side we find large systems in definite states; on the other side we find small systems in indefinite states. But as we saw in the case of thermodynamics and Brownian motion, there is no clear division between 'large' and 'small'.

Imagine an experimenter with their equipment getting smaller and smaller. In practice this could be achieved by replacing the experimenter with a machine, and considering machines operating at smaller and smaller scales.[3] At *what point* does the experimenter transition from a definite state of reality to an indefinite state of unreality? Serious attempts to answer this question are bound to fail and will only raise more unanswerable questions.

The experimenter's apparatus is made of atoms. How can the apparatus be in a definite state when its constituents are not? How can something real and definite be composed of something unreal and indefinite? It is often argued that, somehow, it all works out just fine when the number of atoms is sufficiently large. But this begs the question of *how many* atoms are required. If we add one atom at a time to a system, gradually making it larger and larger, do we reach a tipping point where the addition of a single atom suddenly makes the system transition from an unreal indefinite state to a real and definite one? That is wholly implausible. The fact is there can be no sharp dividing line and the distinction at the heart of quantum mechanics is imprecise and bogus.

The more we think about quantum mechanics, the less sense it makes. Consider again Schrödinger's unfortunate cat. If an experimenter actually looks at the cat, they will find the cat to be in one state or the other—that is, either dead or alive. But before the observation, the cat is in some sort of limbo in which it is both dead and alive. The effect of an experimenter's observation seems profound indeed, causing the cat to collapse from an unreal indefinite state to a real and definite one. But now there is a question: what if we replace the experimenter by a mouse or a fly—or a camera (Figure 13)? Does collapse to a definite state occur only when a human being looks at the cat, or will an animal or an insect do?

---

[3] Nanoscientists have already built machines that operate on the molecular scale so this is not entirely fanciful.

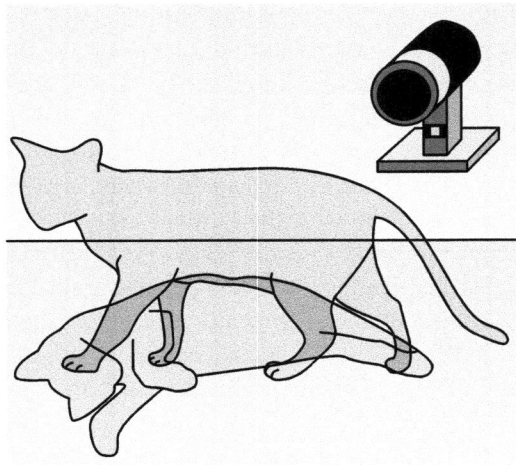

Figure 13. Who or what is an 'observer'? A camera takes a photograph of Schrödinger's cat. Does this make the cat collapse to a definite (alive or dead) state? Or does the collapse happen only when a human observer looks at the cat—or at the photograph?

Or a machine? These questions have no clear answer in standard quantum mechanics.

Here is an example of the kind of maddening confusion people are confronted with when they try to make honest sense of the theory. An experimenter is in the laboratory watching an atom—which may, or may not, decay. Someone else is outside the window, looking in at the experimenter (Figure 14). Does collapse to a definite state occur when the experimenter looks at the atom—or only when the person outside looks at the experimenter? From the point of view of the person outside, if the atom is in an indefinite state the watching experimenter will transition to an indefinite state as well. Like Schrödinger's cat, the experimenter ends up in a kind of limbo: he sees the atom both decay and not decay. The experimenter is literally *in two minds* as to what he has seen. It is only when the person outside makes a measurement—for example, asks the experimenter what he saw—that the experimenter's mind collapses to a definite state (he saw a decay, or he did not). This bizarre scenario was first considered by the Hungarian-American theorist Eugene Wigner in 1961. Wigner rightly concluded that the standard account was absurd. Do we really believe that other people's minds have no definite content until we ask them? What

Figure 14. Who observes the observer? An atom is observed by an experimenter, who is in turn observed by someone outside the window. Is the experimenter's mind in a fuzzy state of quantum limbo—until someone outside asks him what he saw?

happens if *you*, the reader, are that experimenter? Is *your* state of mind indefinite until someone asks you about it?

Despite ongoing attempts to play it down, the sense of confusion surrounding quantum mechanics continues to increase, with critical voices growing louder. The awkward truth—ever more openly acknowledged but still widely dismissed—is that the quantum mechanics taught in our universities and in our textbooks is incoherent. It does not make sense. It is at best imprecise and at worst obscurantist. Inevitably, one day, it will be replaced by a theory that *does* make sense.

The key difficulty is this. Where exactly is the dividing line between the fuzzy unreal world of microscopic atoms and the definite macroscopic reality of our everyday senses? There is and can be no precise answer to this question, which is ultimately why the theory does not make sense. The only way forward is to recover a sense of reality and come up with a precise account of what is actually happening inside a quantum system—even when there is no experimenter around to observe it.

But now is not the time?

Today only a small number of physicists worldwide are actively pursuing research into quantum reality. For nearly a century, students have been warned against wasting their time on such work. Sharp questions about reality in quantum mechanics continue to be widely dismissed.

Even those who accept that there is something wrong with quantum mechanics often remark—with an air of well-informed judgement—that 'the time is not ripe' to tackle these questions. One day something will be done about it, but now is not the time. We do not know enough. Let us get on with other things: developing quantum computers, completing our theory of particle physics, understanding black holes and quantum gravity. As our knowledge advances, we will eventually stumble on some clue that will help us understand quantum mechanics itself. After all—so the reasoning goes—there have been many times in the history of science when people tried to understand something prematurely. Let us not make the same mistake.

This attitude seems unfounded and defeatist. For all we know, a breakthrough could be just around the corner. If we do not look for it, how will we know? The answer might be staring us in the face—and no one sees it because so few work on it because hardly anyone believes that now could be the time.

It might be pointed out that there are reasons why, for example, hard evidence for atoms and molecules did not emerge until the early twentieth century. How could scientists in, say, the early nineteenth century have possibly come up with something like kinetic theory and its explanation for Brownian motion? Sometimes we just have to be patient and wait for the right time.

In fact, in a book published as long ago as 1738—in the early *eighteenth* century—the Swiss physicist and mathematician Daniel Bernoulli outlined what we now call the kinetic theory of gases: among other things he showed that atomic or molecular impacts could explain the pressure of a gas (Figure 1). Bernoulli was, as they say, ahead of his time. His ideas were ignored. A full century later, English amateur scientist John Herapath and Scottish engineer John Waterston developed similar ideas—but their papers were rejected for publication by the prestigious Royal Society of London. Such ideas only began to be taken seriously in 1857, when a German physicist by the name of Rudolf Clausius published an influential paper about kinetic theory. For several decades thereafter, kinetic theory was championed and developed in particular by Boltzmann and also by the great Scottish physicist James Clerk Maxwell. Eventually, in 1905, Einstein showed precisely how atomic or molecular impacts could explain the microscopic jittering which had been seen by Brown some eighty years earlier.

How ironic. Evidence for the existence of atoms and molecules was there under the botanist's microscope for all to see as early as 1827, around the time when Herapath (in 1820) and Waterston (in 1845) were rediscovering the essential ideas required to understand it—ideas that had first been put forward by Bernoulli nearly a century before. Why did it take until 1905 to understand that the jittering motion seen by Brown was caused by impacts of invisible water molecules? After all, the essential idea is simple and natural enough. We see a tiny speck that is continually shaking. Perhaps it is being knocked around by even smaller specks we cannot see. This idea is so simple and natural, in fact, that it was outlined more than two thousand years ago. In the first century BCE, in his book *On the Nature of Things*, the Roman poet and philosopher Lucretius argued that the jittering motion of fine particles of dust floating in sunlight was caused by the impacts of invisible atoms.

So why did we have to wait until 1905 for a full understanding of Brownian motion? Of course some weighty technicalities had to be worked out in order for Einstein's calculations to succeed. But it is difficult to avoid the conclusion that progress was also unnecessarily retarded, firstly by a kind of intellectual timidity that saw Bernoulli, Herapath, and Waterston as just too speculative, and secondly by the widespread philosophical opposition which emerged in the late nineteenth century—from Mach and others, who insisted that a proper physics should speak only of sensory perceptions and not posit imaginary things we cannot see.

All that may sound like an extraordinary historical exception. Surely, in the case of Ptolemaic astronomy for example, we cannot possibly argue that the ancient Greeks might have formulated a theory of astronomy with the planets orbiting the Sun? There are some things for which, it has to be admitted, the time is simply not ripe.

Well, in fact, in ancient Greece someone *did* construct a heliocentric model of the world—and long before the time of Ptolemy. In the third century BCE, Aristarchus of Samos proposed that the Sun is at rest, with the earth and other planets revolving around it. He also understood that the stars were other suns located much further away. But Aristarchus' model was rejected in favour of the Ptolemaic system. Why?

It might be thought that Copernicus was able to reverse the error only because he knew more. But the astronomical data and mathematical methods available to him were essentially the same as those available to the ancients. It is difficult to avoid the conclusion that, again, a combination of intellectual timidity and philosophical preferences played a role in supporting the Ptolemaic system—with its unbridgeable chasm between our mortal earthly understanding and the eternal ethereal heavens.

Modern scientists sometimes hold naïve views about history and scientific progress. In the past science has been lost down a blind alley for decades or even centuries, with the simple truth being pointed out in vain by neglected pioneers. There is no reason to think it will not happen again.

Louis de Broglie's 1927 pilot-wave theory

According to conventional wisdom, the old-fashioned idea of an objective physical reality was decisively overthrown by the rise of quantum mechanics in the 1920s. It therefore seems inconceivable that anyone a century ago could have concocted a working model of quantum reality. After all, our best physicists are unable to accomplish such a task even today, so a century ago it must have been completely hopeless.

As it turns out, history repeated itself yet again. In 1927 a working model of quantum reality *was* successfully constructed—by the French physicist Louis de Broglie.[4] He called it 'pilot-wave theory'. But de Broglie's ideas were rejected and soon forgotten. In 1952 pilot-wave theory was revived and extended by the American physicist David Bohm. The theory was again rejected and forgotten. Like the pioneers of kinetic theory, de Broglie and Bohm were either ahead of their time or at variance with the philosophical preferences of their peers (Figure 15).

According to pilot-wave theory, at the atomic scale there are real particles moving around in space. Each particle is guided by a wave—the 'pilot wave' (Figure 16). Mathematically speaking, the pilot wave is the same as the usual quantum wave (or 'wave function') described in textbooks. But physically it is quite different. The usual quantum wave is merely a

---

[4] Approximate English pronunciation: 'de Broy'.

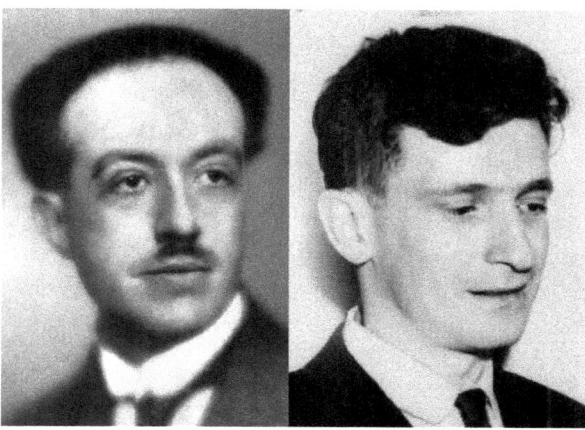

Figure 15.   Neglected pioneers of quantum reality. De Broglie (left) and Bohm (right).

tool for calculating statistics, whereas the pilot wave guides the motion of each individual particle. As a rough analogy we might think of a bottle floating on the surface of the ocean. The motion of the bottle depends on the wave that carries it around. If we throw a bottle into the water, then depending on the ocean currents it will follow a definite trajectory, and eventually be washed up on a beach somewhere. Obviously, where the bottle ends up depends on where it starts and on the wave that carries it. According to de Broglie, something like this happens to a subatomic particle such as an electron. It does not just pop into existence when observed, nor does it jump around at random. It follows a definite trajectory, moving from one place to another in a way that is determined by its pilot wave. And again, where the particle ends up depends on where it starts and on the wave that carries it. Despite superficial appearances, in the quantum world everything happens for a reason, and in ways that are precisely defined.

We know that pilot-wave theory can explain all of the strange phenomena of quantum physics. Any doubts about that were laid to rest by Bohm in 1952. And yet the theory has only a small number of supporters, and continues to meet with widespread opposition or indifference. Why? As with the kinetic theory explanation of gas pressure in the nineteenth century, it can be difficult to understand why so many scientists are

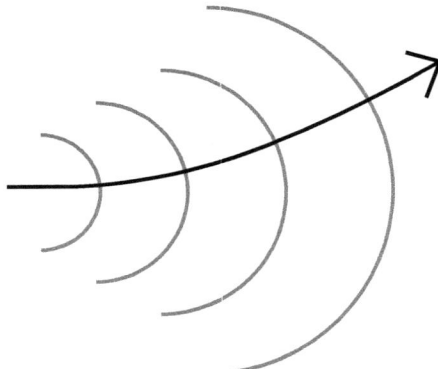

Figure 16. De Broglie's pilot-wave theory. A moving particle (dark line) is guided by a wave.

opposed to it or at best indifferent. In part, there are the usual philosoph-ical reasons. As in previous ages, in our time perception has once again come to be preferred over reality. The underlying trajectories described by de Broglie and Bohm cannot be seen directly or in detail (at least not at present). As was the case with kinetic theory and thermodynamics, there is a widespread prejudice against invisible explanations for what we can vis-ibly see. Pilot-wave theory also carries a theoretical price tag which most physicists are unwilling to pay: widely separated particles can affect each other instantaneously, no matter how far apart they may be—in conflict with Einstein's relativity, which forbids superluminal or faster-than-light influences. At the deeper level described by pilot-wave theory, Einstein's relativity must fail. For many physicists this is unacceptable. It appears that, for some, no reality at all is preferable to a reality without the theory of relativity.

Thinking through the implications of pilot-wave theory can lead to some unsettling conclusions. As we will show in this book, the theory is not just an alternative account of quantum physics. It is much more. In certain con-ditions, pilot-wave theory agrees with the laws of physics as we currently know them. But in other conditions, the theory predicts a radical break-down of those laws. Both quantum physics and relativity turn out to be valid only in restricted circumstances—which we happen to be confined to. Both theories fail more generally, as this book will explain. This means that, at present, our most advanced scientific experiments are seeing only fragments of what is possible.

According to pilot-wave theory, at least as understood in this book, there is a whole new physics 'beyond the quantum'. In this new physics we can see quantum reality clearly and bring it under precise control. It then becomes possible to signal instantaneously across space, in blatant violation of accepted physical laws. Behind the incomprehensible mathematical symbolism of quantum mechanics, there lies hidden a physical reality unlike anything we have ever seen. And, at least in principle, it is possible for us to access and control that deeper reality.

Needless to say, extraordinary claims require extraordinary evidence, and until we find concrete experimental evidence for the wider physics described in this book, most physicists will remain sceptical. That is understandable, and indeed scientifically sound. In the end, scientific truth can only be decided by experiment, and at the time of writing there is no firm evidence for an experimental breakdown of quantum physics. This book will argue, however, that there are multiple ways in which quantum physics not only could break down but (probably) *should* break down—and there are myriad ways in which these predictions and expectations can be tested experimentally.

Historically, however, most supporters of pilot-wave theory have turned a blind eye to the implied new physics beyond quantum mechanics. In their view, the theory can do no more than provide a clearer explanation for the same physics we already know, and there is no reason to think that the usual predictions of quantum mechanics will ever break down. It seems that, even for those physicists who are sympathetic to the cause, what pilot-wave theory actually has to say about our world—and our place in it—is too troubling to contemplate. As we will see, on the contrary, pilot-wave theory is interesting precisely because it naturally implies the possibility of new physics beyond what is presently known.

What this book is about

Pilot-wave theory tells us that, beyond the quantum physics we see around us today, there is a deeper reality in which all things are instantaneously connected right across the universe. But, if such a deeper reality truly exists, why can we not see it directly here and now? This is a good question.

As we will discuss in detail, the deeper reality is hidden from us because our world is permeated by a sort of 'quantum fog'—a universal background noise which is, in some respects, analogous to the incessant jittering seen in Brownian motion (Figure 11). The fog prevents us from seeing the underlying reality clearly. The noise makes it impossible to control. As a result, the quantum world we now see appears to be random and uncertain. Behind the facade, however, reality is profoundly different. Like ancient Ptolemaic astronomy, quantum mechanics is supremely successful as a tool for making calculations. But it misses the deeper reality that is now hidden from us.

According to quantum theory, as usually taught, 'quantum noise' is an unavoidable part of the fabric of reality. It is dictated by the laws of quantum mechanics. There is one particular law, or formula, which physicists call the 'Born rule'. It is named after the German physicist Max Born, who proposed it in 1926. The formula describes the quantum noise that permeates everything in our world.[5]

Because of the all-pervading quantum noise, we cannot measure the trajectories of quantum particles accurately. More specifically, we cannot measure both the position and the velocity of a particle at the same time. This limitation is called the 'uncertainty principle'. It was first proposed by Heisenberg in 1927. The uncertainty principle is widely regarded as having finally demolished naïve ideas about reality: if particle trajectories cannot be observed, then they do not exist. The particle trajectories calculated by de Broglie and Bohm are metaphysical fictions—much as Boltzmann's bouncing atoms and molecules had been for Mach.

But pilot-wave theory—as interpreted in this book—tells us that Born's formula and Heisenberg's uncertainty principle are not immutable laws of physics, as is widely assumed, but are instead merely features of a special state which we happen to be confined to. In this state everything is permeated by the same obscuring fog or noise, which prevents us from making accurate measurements. But the world does not *have* to be this way. Nor is there any reason to believe that the world always was this way.

The quantum fog we see in our world today probably formed near the beginning of the universe. On this view, the jittering or noise described by Born's formula was created by the violence of the 'Big Bang' (the gigantic

---

[5] Technically, as we will see later, Born's formula states that quantum probabilities are given by the squared magnitude of the quantum wave.

explosion at the beginning). To catch a glimpse of reality as it is, we can try peering back to those early times. The 'cosmic microwave background'—a much-studied radiation left over from the Big Bang—in effect provides us with a fossil record of the beginning. It may contain vital clues about what the universe was like before the fog took over.

Is it possible once and for all to free ourselves from the obscurantism which Born's formula and the uncertainty principle have consigned us to? Might we see quantum reality for what it truly is? This book argues that it should indeed be possible.

Relic particles from the early universe are widely thought to exist in our world. They may well be a component of the 'dark matter' that is being avidly searched for by astronomers. Perhaps some of these particles managed to escape the violence of the Big Bang before the quantum fog took over. If we can find such untainted particles, we will have discovered a radically new form of matter—'subquantum matter'—that violates the laws of quantum mechanics. Born's formula would be wrong. With the aid of subquantum matter we could beat the uncertainty principle. We could see through the fog and control the underlying reality directly.

It is also conceivable that the obscuring fog could be torn apart by exotic effects involving black holes and quantum gravity. According to these ideas, gravity can render quantum mechanics 'unstable': if Born's formula is true today, it need not be true tomorrow. It may then be possible to *create* subquantum matter where there was none before.

If it were discovered or created, subquantum matter would provide us with a window onto quantum reality. It could be deployed to communicate faster than the speed of light—instantaneously, in fact, even over vast distances. This would violate Einstein's theory of relativity, which precludes superluminal communication. Time would no longer be 'relative': instantaneous signals would define a true time across the universe. Subquantum matter would also allow us to crack quantum codes, and to perform many other seemingly impossible tasks. Such a discovery would no doubt herald a technological revolution whose outlines we can only dimly foresee.

According to pilot-wave theory, as we understand it in this book, the quantum physics we see around us today is only a special case of a wider and richer physics, which probably existed in the remote past and which may still be accessible now. Today our ability to see and control quantum reality is severely limited by the obscuring effects of quantum noise. The textbooks say that these limitations are unbreakable laws of physics.

Pilot-wave theory says no. We should be able to find our way out of the fog and beat the uncertainty principle. If we are successful, we will unveil a deeper reality in which everything in the universe is instantaneously connected with everything else. We will finally see quantum reality for what it is. A whole new physics will be opened up to us.

Our story begins where it is usually thought to end: in the 1920s, when according to received opinion the meaning of quantum mechanics was finally settled and the idea of reality was relegated to a quaint relic of the nineteenth century. As we shall see, nothing could be further from the truth.

1

Before the Quantum

A theory of quantum reality—and its rejection, twice

Louis de Broglie spent the 1920s working in relative isolation in Paris. Few understood what he was doing or why. To this day, many physicists and historians have a peculiar blind spot for his work. De Broglie's aim was nothing less than a complete overhaul of the foundations of physics. He set out to construct a new theory of motion—a radical break from the laws of motion as given by Newton and Einstein—which would work at the atomic scale and explain the mysteries that were pouring out of the laboratories of Europe.

By 1927 de Broglie had succeeded. He was ready to present his pilot-wave theory to his peers. He did so at what is probably the most dramatic, significant—and profoundly misunderstood—scientific conference ever held in the history of physics: the 1927 Solvay conference (Figure 17).

The conference was held in Brussels and attended by the foremost physicists of the day. It is usually remembered for a debate between Bohr and Einstein, in which Einstein supposedly tried and failed to circumvent the uncertainty principle. It is widely believed that this debate helped establish a consensus about the meaning of quantum mechanics. In fact, there was no 'debate' between Bohr and Einstein, in the sense of a moderated public discussion. There were instead some lively conversations between the two men, mainly over breakfast and dinner, which were overheard only by Heisenberg and one or two others. And even in those conversations, Einstein's point was misunderstood. For Einstein's concern was not the uncertainty principle, but rather what he regarded as the 'spooky' character of quantum theory (about which we will have much to say). Our collective historical perception of what took place at the conference is badly distorted, mainly by relying on misleading accounts written decades later by Bohr

Beyond the Quantum. Antony Valentini, Oxford University Press. © Antony Valentini (2025). DOI: 10.1093/oso/9780198853749.003.0002

Figure 17. Louis de Broglie (circled) presented his pilot-wave theory at the famous—
though widely misunderstood—1927 Solvay conference.

and Heisenberg, who gave the impression that those semi-private conversations with Einstein had been central to the conference when, in fact, they
were quite peripheral.

One of the highlights of the conference was de Broglie's lecture, where
he presented his pilot-wave theory. It triggered lively discussion and extensive debate. The other main highlights were a lecture by Schrödinger
and a lecture by Born and Heisenberg. Far from establishing a consensus, the truth is that *three* radically different and rival interpretations of
quantum physics were presented and discussed—with *no* resolution or
agreement.

Most historians have missed what the conference was really about. The
conference proceedings were published in French in 1928. They contain a detailed record not only of the lectures but also of the extensive
discussions among the participants. The published discussions contain
no trace of the famous but quasi-mythical 'Bohr–Einstein debate' about
the uncertainty principle. For the rest of the twentieth century, the proceedings languished in dark corners of university libraries, unread and
misunderstood. A complete English translation, with detailed analysis and

commentary, was published in 2009.[6] There were many surprises. Remarkably, none of the three interpretations presented bore a close resemblance to what we now call quantum mechanics. But that is another story. Of special interest here is the quality and completeness of de Broglie's presentation.

In de Broglie's 1927 pilot-wave theory, each particle is guided by a wave—rather like a floating bottle carried along by an ocean wave (Figure 16). But de Broglie's pilot wave is more abstract than an ocean wave. The law by which the wave makes the particle move is unlike anything in Newton's or Einstein's physics. Instead of a force causing acceleration as envisaged by Newton (Figure 6), in de Broglie's theory we have a pilot wave causing *velocity*. This is reminiscent of Aristotle's ancient belief that even a constant velocity has to be caused by something. In short, pilot-wave theory is a radically different way of thinking about motion. Because of the complexity of the wave, the particle can follow bizarre wriggling trajectories which are impossible in Newton's (or Einstein's) physics. This novelty is what enables de Broglie's theory to explain quantum physics.

De Broglie's theory was, as we have said, the subject of long and lively debates at the Solvay conference. But it was soon forgotten, and eventually de Broglie himself abandoned it. In 1929 de Broglie won the Nobel Prize in Physics for being the first to propose that material particles could behave like waves. But the pilot-wave theory that had led de Broglie to this startling conclusion was disregarded.

De Broglie's theory was revived in 1952 by David Bohm. In some respects Bohm misunderstood the theory. In particular, he did not understand de Broglie's radical departure from Newton's laws. But in other ways Bohm took the theory further. Most importantly, Bohm showed in detail how pilot-wave theory is able to reproduce all of the usual predictions of quantum mechanics.

Like de Broglie's original theory, Bohm's work was again quickly forgotten. Bohm had held a prestigious position at Princeton University, which he lost for political reasons. He then worked in Brazil for a few years and, after a stint in Israel, eventually settled in London—where for the rest of his life he lived out a kind of exile from mainstream physics.

---

[6] This work was carried out by the author in collaboration with the Italian philosopher of physics Guido Bacciagaluppi.

Thirty years ago most physicists still believed that a theory of quantum reality was impossible. Hardly anyone had read what de Broglie and Bohm had written on the subject. Pilot-wave theory, if it was considered at all, was usually dismissed as definitively disproven. Such misunderstandings and misinformation are less common today. Even so, there is one feature of pilot-wave theory that still provokes widespread incredulity.

Nonlocality: or spooky action at a distance

Pilot-wave theory has a remarkable—in fact, amazing—feature which was not properly understood until much later. Two particles that are far apart can push and pull on each other instantaneously. And it does not matter how far apart they are. One could be here in the lab and the other a million light years away, it would make no difference. They can be directly connected *instantaneously*.

This happens when the particles are 'entangled'—which essentially means there is one pilot wave guiding both of them together. This will be explained properly in Chapter 2. But to get a rough idea for now: strange as it may sound, for two particles the pilot wave really exists in a higher-dimensional 'configuration space' that contains information about both particles. The action of the wave then makes the motion of one particle depend instantaneously on the other particle—no matter how far apart they may be (Figure 18).

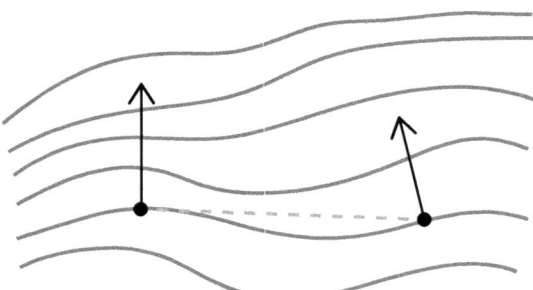

Figure 18. Spooky action at a distance. Two entangled particles are guided together by one pilot wave. The particles are connected: the motion of one depends instantaneously on the other. (This simplified figure is only schematic. For two particles the pilot wave really exists in a higher-dimensional space. See Chapter 2, Figure 39, for a more accurate picture.)

Quantum mechanics has much to say about 'entanglement' between distant particles. The theory says nothing about moving particles, however. The pilot wave is called the 'wave function'. It is regarded only as a mathematical tool to calculate probabilities (or statistics). So what does quantum mechanics say if two particles are entangled? As usual, the textbooks are not very clear. When two particles are entangled, their joint quantum wave (or wave function) cannot be separated into individual waves, one for each particle. As a result they *appear* to have a direct connection. But in practice no real signal from one to the other is discernible. It is as if the particles somehow know about each other, but cannot really communicate. In pilot-wave theory it is more explicit: entangled particles really do communicate, instantaneously, no matter how far apart they are. There is no hand waving or equivocation: it is not 'as if' the particles are connected—the particles *really are* connected.

This kind of instantaneous connection is often called 'nonlocality' or action at a distance. In conventional physics an effect here can only come from 'local action', which means that it is caused by something nearby and not directly by something far away. Einstein's theory of relativity—formulated in Einstein's miracle year of 1905—teaches that no physical influence can travel faster than the speed of light. If we want to send a signal from here to there, the best we can do is send a light (or radio) signal. To signal faster is against the laws of physics—or at least, against those laws as they were written by Einstein.

Trying to break the speed-of-light barrier is bad enough. To posit physical influences travelling *infinitely* fast flies in the face of everything twentieth-century physicists thought they had learned. For this reason, nonlocality is often called 'spooky'. Famously, Einstein dismissed such effects as impossible.

A strange conspiracy

It seems that pilot-wave theory flagrantly disregards the most basic principles of twentieth-century physics. In fact, it is more subtle than that. Pilot-wave theory tells us that particles follow strange wriggling trajectories. But in practice the details of those trajectories can never be seen or measured. The theory also tells us that a particle here can respond instantaneously to a particle far away over there (if the particles are entangled). But in practice the details of that response can never be seen or measured. The burning question is: why *hide* all this?

We can put it another way. If the particle trajectories conjured up by de Broglie and Bohm really exist, why can we not see them? If particles here in the lab really are in instantaneous communication with particles far away, why can we not make use of this to send instantaneous messages across space? It is as if there is some sort of conspiracy. Something is going on behind the scenes—something which we are never allowed to witness or control. Why?

In short: if pilot-wave theory is right, why can we not see and control the remarkable quantum reality it describes? For the answer, we must first step back to a truly grim and fearful image of our future which was first forecast in the nineteenth century.

Heat death of the universe: or what happens when the lights go out

Despite what many ancients believed, the night sky is not eternal. Up there fires are burning—nuclear fires, but fires nonetheless. They are called stars. One of them is so close that during the day you can feel its heat on your skin: we call it the Sun. Without heat and light from the Sun, life on Earth would perish. The other stars are much further away. Many can be seen only with powerful telescopes. Each of these fires, including the Sun, will eventually burn out (in billions of years, Figure 19). The lights will eventually go out. All will be darkness and cold. What then?

Figure 19. Heat death of the universe. Today the Sun and stars are burning bright (left). In the far future they will start to burn out (middle), eventually leaving darkness and cold (right). (We are simplifying a bit. We now know that stars have a complex life cycle, which can end in an explosion or supernova. But the remnants will cool and the overall conclusion is correct.)

This stuff of nightmares originated in the 1850s and 1860s. Even though what powered the stars was not yet known, physicists were able to reason from the general principles of thermodynamics. It was shown that, in a precise sense, the universe is running down and will eventually die. Thermodynamics describes how heat can be converted into work: hot steam, for example, can be harnessed to drive an engine. But to do this we need differences of temperature: if this is hot and that is cold, heat will flow from hot to cold and we can extract what is called 'mechanical work'. But thermodynamics also describes how temperature differences eventually even out and disappear. Hot things cool down, cold things warm up, and eventually everything will reach the same temperature. This end state came to be known as the 'heat death of the universe'. And for good reason: once this state is reached, it is no longer possible to convert heat into work. That may not sound so drastic—to those unfamiliar with thermodynamics. But if heat cannot be converted into work, all life and activity as we know it comes to an end. And perhaps worst of all: once the end state is reached, there is no way out. We are stuck there, essentially forever.[7]

What has this dismal future got to do with quantum mechanics? You may well be shocked by the answer.

Quantum death of the universe: why reality is invisible

This book argues that quantum reality is invisible to us for a simple but profound reason. Something like the feared heat death of the universe *has already happened*. At the quantum level, our universe is dead. Figuratively speaking, the lights have already gone out and we are unable to see the world as it really is. The lights probably went out a long time ago—at the beginning of our universe, soon after the Big Bang. We call this the 'quantum death of the universe'.

That is why our universe is permeated by a fog of quantum noise which prevents us from seeing reality clearly. Physicists talk about Heisenberg's uncertainty principle. But the fog and uncertainty that surround us today are not set in stone by the laws of nature; instead, they probably formed soon after the Big Bang. Before that time, there was no uncertainty principle and

---

[7] In principle, an extremely rare 'thermal fluctuation' could by chance drive the universe away from heat death, but in practice this is so statistically unlikely as to be of no consequence.

our universe was suffused with superluminal signals. At the beginning it was possible to see and control reality at the deepest level. Fast forward billions of years to the present: we are now surrounded by a background quantum noise that frustrates our attempts to see and control reality, and which prevents us from signalling faster than light.

Despite the obscuring fog, some physicists have been astute enough to understand that behind the scenes our universe is pervaded by spooky action at a distance. But we cannot control or make use of it. In the state of quantum death, the deepest level of reality remains invisible and uncontrollable. This is not because of any 'conspiracy'. It is simply a peculiarity of the state of quantum death which we happen to be confined to. The uncertainty principle and the speed-of-light signalling barrier are not laws of physics but mere features of quantum death. In an earlier state of 'quantum life', these restrictions do not apply.

We have some explaining to do. Let us look a bit more closely at the heat death described in the nineteenth century. Imagine what it would be like to live in a world where everything has reached the same temperature. Life could not exist there, but for the sake of argument we can just imagine what hypothetical beings in such a world might experience (Figure 20). Steam engines could not work. And generally, without differences of temperature to set up a heat flow, heat energy could not be employed for anything useful. Scientists living in such a 'zombie universe' would think that this limitation is a law of physics. Perhaps the law would be named after someone: according to so-and-so's law, it is impossible to convert heat energy into useful mechanical work.

There is more. Everything in the heat-death universe would be jittering like the tiny granules under Robert Brown's microscope (Figure 11). Intriguingly, because the temperature is everywhere the same, the *amount* of jittering would be the same—whether we were talking about molecules, granules, or the piston of a steam engine, all would be subject to the same (tiny) jittering motion. The technical term is 'thermal noise'. Everything in the universe would be bathed in the same thermal noise—a kind of inescapable background hiss that pervades all things. Again, physicists in such a world would come up with 'so-and-so's principle': it is a fundamental law of nature that every system in the universe has the same unavoidable jittering or thermal noise.

Now imagine what would happen if an experimenter living in the heat-death universe tried to see and control the motion of individual molecules. Everything is shaking or jittering. The molecules which the experimenter is trying to see are shaking. The experimenter's microscope is shaking (a little bit). In fact all of the laboratory equipment is shaking. Even the experimenter's own hands and eyes are shaking. And the (tiny) amount of shaking, or jittering, is always the same. So imagine your hand is shaking, and you are trying to pick up something which is also shaking—and that your hand is shaking as much as the thing you are trying to pick up—then, well, you can see the difficulty. If we were confined to such conditions, no doubt someone would formulate a 'thermal uncertainty principle': it is impossible to see and control the motion of molecules to an accuracy better than the thermal noise that permeates all things. And finally, in such a world, someone, somewhere, of a certain kind of philosophical disposition, may well come to a truly stunning conclusion: because the finer details of the universe are forever unobservable, those finer details do not really exist.

If we could communicate with beings in such a world, what would we tell them? We would surely want to tell them: *no*, the limitations you experience, and the jittering or thermal noise you find everywhere, and

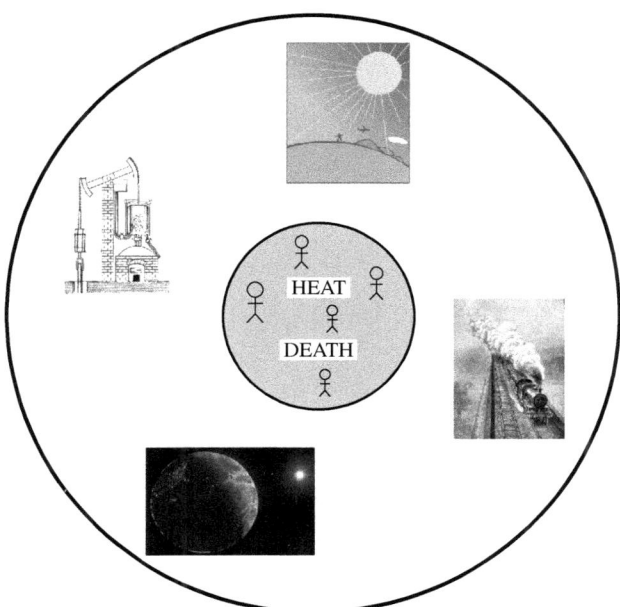

Figure 20. The heat 'deathtrap'. Imaginary beings in the grey circle inhabit the heat death: a world in which everything has reached the same temperature. Outside there is a new and unexplored continent of sunshine, steam engines, and life.

your inability to see and control the finer details of reality, are not laws of nature but mere accidents of the state of heat death which you happen to be confined to. There is a whole new physics out there which you have never dreamed of: heat flow can drive engines to power the world, and it is possible to see and control single molecules. At the deepest level, reality does exist. But to see all this you will somehow have to find your way out of the heat deathtrap (Figure 20).

This book tries to convey a similar message to people living in our world. We are unable to see quantum reality clearly. We cannot see or properly control the motion of subatomic particles. Because of this, we cannot harness entangled particles for practical instantaneous signalling. Everywhere we look, Heisenberg's uncertainty principle gets in our way. All things are permeated by a universal and seemingly unavoidable jittering or quantum noise. We believe that this is part of the fundamental fabric of the universe. Scientists tell us that these limitations are unbreakable laws of nature. But the message of this book is *no*, the limitations we experience,

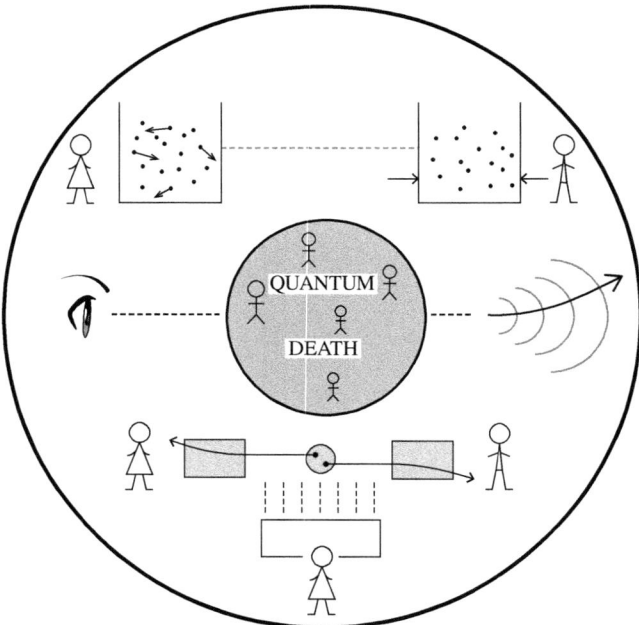

Figure 21. The quantum 'deathtrap'. Human beings in the grey circle inhabit the quantum death: a world in which everything is infused with the same quantum noise or uncertainty. Outside there is a new and unexplored continent of instantaneous signalling, direct manipulation of quantum reality, and radical new technology.

and the quantum jittering or noise we find everywhere, and our inability to see and properly control quantum reality, are not laws of nature but mere accidents of the state of quantum death which we happen to be confined to. There is a whole new physics out there which we have never dreamed of: it is possible to see and control the precise motions of entangled sub-atomic particles, which can then be harnessed to signal instantaneously across space. At the deepest level, quantum reality does exist. But to see all this we will somehow have to find our way out of the quantum deathtrap (Figure 21).

Quantum relaxation: how quantum death happens

Well, that is quite a claim. But what exactly is quantum death and how does it happen? We will be talking a lot about that later in this book. But to get a rough idea for now, let us first think a bit about how the conventional heat death happens.

How exactly do differences of temperature even out? For example, if we put an ice cube on a table at room temperature, what makes the ice warm up and melt? The answer is the complicated knocking around of atoms and molecules—of the sort that causes pressure. Hotter molecules move faster and knock harder against the slower and colder molecules, making the latter move faster and so become hotter (Figure 22). This atomic and molecular knocking around is ultimately responsible for what we call 'heat flow'. Technically, this process is called 'thermal relaxation'. The end state

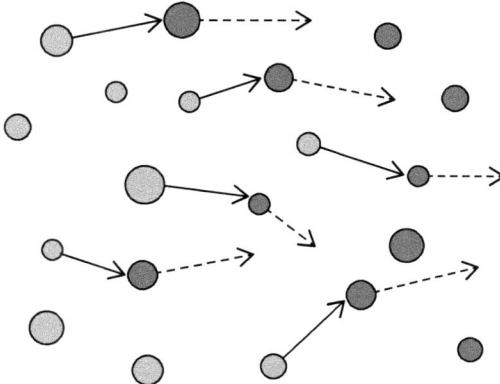

Figure 22.   What happens during thermal relaxation. Hotter and faster molecules (light grey) collide with cooler and slower molecules (dark grey). The hotter molecules slow down and the cooler molecules speed up.

is called 'thermal equilibrium'—the temperature has evened out, and once we get there we stay there.

Something broadly analogous happens in pilot-wave theory. We have said that de Broglie's particles tend to follow peculiar wriggling trajectories. An example is shown in Figure 23, where we track the motion of a particle trapped inside a box. The motion has been calculated with de Broglie's theory.[8] In Newton's physics the particle would bounce off the walls, and between bounces it would move in simple straight lines (figure inset). But in de Broglie's physics the particle follows a complicated wriggling motion. Detailed calculations show that this wriggling motion is precisely what is needed to move from quantum life towards quantum death. For large numbers of wriggling particles, we eventually end up in a state where, roughly speaking, all the particles are pervaded by the same quantum noise.[9] Technically, this process is called 'quantum relaxation'. The end state is called 'quantum equilibrium'—once we get there, we stay there (Figure 24).

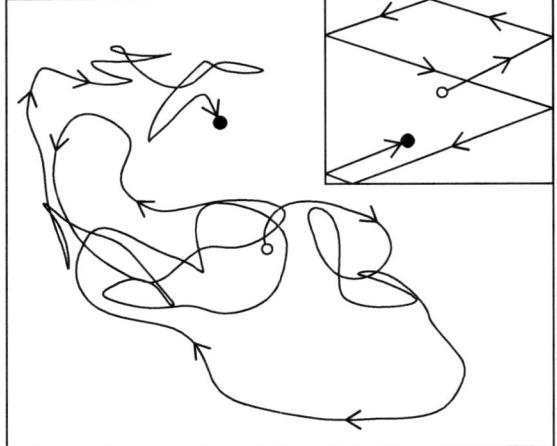

Figure 23.   What happens during quantum relaxation. In de Broglie's physics a particle trapped inside a box has a strange wriggling motion, beginning in the small white circle and ending in the small black circle. In Newton's physics, in contrast, the particle bounces off the walls and moves in straight lines between bounces (inset).

[8] Technically, the pilot wave for this example is a superposition of 16 energy states.
[9] We are being a bit loose here. For a proper discussion see Chapter 5.

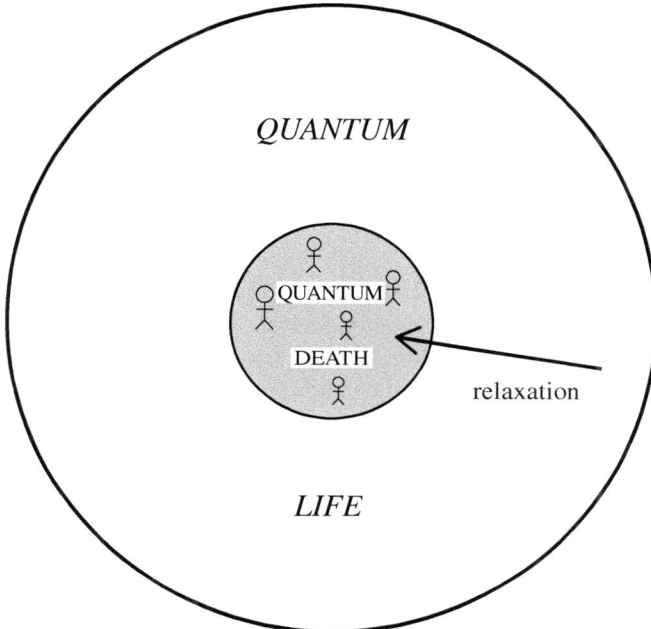

Figure 24.   From quantum life to quantum death, or relaxation to quantum equilibrium. Born's formula and quantum noise apply *only* to the inner 'death' zone.

In the case of heat death, there is a formula that describes the jittering or thermal noise of the end state. It was first worked out by Maxwell in 1859. Physicists call it the 'distribution of molecular speeds'. Maxwell's formula tells us how fast molecules are likely to be moving when they have reached a specific temperature. That formula is of course not a fundamental law: it applies *only* to the special end state (thermal equilibrium). More generally—in what is called a state of 'thermal nonequilibrium'—molecules are not distributed according to Maxwell's formula.

Similarly, in pilot-wave theory, there is a formula that describes the jittering or quantum noise of the end state. This formula turns out to be the same as one of the formulas already found in quantum theory: the 'Born probability rule' or Born's formula, first proposed in 1926, and which we met briefly in the Prologue. The formula states that the probability to find a particle somewhere is given by the squared magnitude of the quantum wave at that point—where the 'magnitude' is simply the height or size of the wave. So the particle is more likely to be found where the wave is larger, and less likely to be found where the wave is smaller. The quantum textbooks tell us that Born's formula is a fundamental law of nature, valid always and everywhere. But in pilot-wave theory Born's formula is not a fundamental law: it applies *only* to the special end state (quantum equilibrium). More generally—in what is called a state of 'quantum nonequilibrium'—particle positions are not distributed according to Born's formula (Figure 24).

Our imaginary beings in the zombie heat-death universe would be sceptical if we told them that Maxwell's formula for molecular speeds is not a universal law but only a special case they happen to be stuck with. Their scepticism would be understandable: heat death is all they have ever known, and their formula successfully describes what they see. Why should they believe anything different? Only if we can show them a way out, so they can experience something beyond heat death and see for themselves.

Similarly, scientists in our world are sceptical when they are told that Born's formula for quantum noise is not a universal law but only a special case they happen to be stuck with. Again their scepticism is understandable: quantum death is all they have ever known, and Born's formula successfully describes what they see. Why should they believe anything different? Only if we can show them a way out, so they can experience something beyond quantum death and see for themselves.

Three ways out of quantum death

So now we have dug ourselves into the deepest of holes. If there is no way out of quantum death, there will never be concrete evidence for these ideas and everything we have said will remain just so much speculation. No one will ever believe us. So if we take these ideas seriously, we have to find a way to escape and see reality directly—as it is, without quantum noise or fog or the uncertainty principle getting in the way. But how?

We can find some clues by thinking a bit more carefully about the fearful heat death described in the nineteenth century. Thermal relaxation can be pretty quick (think of throwing an ice cube into a fire). But it is not instantaneous. Similarly, quantum relaxation will take some time. The quantum noise or fog described by Born's formula will not emerge instantly but will take time to form. The question is: when did quantum relaxation happen and how long did it take? Could there be some traces left over—a kind of fossil record—of what existed before?

Let us think about the beginning of our world. Cosmologists generally agree that the universe we see began about 13.8 billion years ago in a hot and violent Big Bang. Physicists usually assume that the laws of quantum mechanics, including Born's formula, have always been true. But in pilot-wave theory Born's formula is only a special case and there is no reason to think that the universe began that way. So let us imagine what would happen if our universe began in a general state *not* described by Born's formula. At the beginning quantum mechanics would be wrong and the universe would be teeming with measurable instantaneous signals. But as we will see in Chapter 5, we can expect the quantum fog to form very quickly—in the earliest moments of the Big Bang.

That settles it, you might think. If quantum death happened very soon after the Big Bang, about 13.8 billion years ago, what hope is there now of finding any traces of what came before? Trying to test this theory is a fool's errand!

But there are three loopholes—or three escape clauses—which make our situation not nearly as hopeless as it might seem.

The first loophole is this. We have said that quantum fog forms very quickly in the earliest moments of the Big Bang. Even so, the universe today might contain 'relic particles' that have survived from those earliest moments—particles that have since been moving freely, untouched by anything else. In fact, cosmologists today often invoke 'dark matter' to explain various astrophysical anomalies. Dark matter is often assumed to be made of relic particles from the very early universe. Now it is possible that some of those particles did not have time to reach the state of quantum death: they could have stopped interacting with other particles before they reached the end state (see Chapter 7).

We might call these particles 'subquantum fossils'. They did not have time to become completely shrouded in quantum fog. Born's formula does not apply to them. They contain *less* noise than ordinary quantum particles. For this reason they can be deployed to circumvent the uncertainty principle, allowing us to see and control reality more accurately (Chapter 5).

If we can find such particles we will have found a radically new form of matter that frees us from the restrictions of quantum death. This 'subquantum matter' would violate the known laws of quantum mechanics. It would have myriad applications beyond those allowed by the restricted quantum technology of today. For example, if subquantum particles were entangled we could use them to send instantaneous messages across space. They could also be deployed to break quantum codes and to build radically new kinds of computers. As with any new technology, it is difficult to foresee where it might lead in the long term. But there is no doubt that the discovery of such matter would herald a new age of 'subquantum technology'. That will be the subject of Chapter 10.

And now for the second loophole. We have said that quantum relaxation—the formation of quantum fog—takes place very quickly in the early universe. Closer analysis shows that this is true only at 'short wavelengths'. Because the early universe is expanding rapidly, the details of relaxation are modified. This gets a bit technical and will be explained in Chapter 6. Roughly, if we study the early universe over small distances we expect the fog to form quickly, but over larger distances the fog will form more slowly. It is then quite possible that, if we perform measurements over large enough distances, we will find a breakdown of Born's formula and a record of the 'pre-quantum' state.

But how could such measurements of the early universe ever be done? We are fortunate to have a snapshot of the early universe ready to hand, at both small and large distances. This snapshot is what cosmologists call the cosmic microwave background or CMB—a relic radiation in space that formed at early times and whose structure provides us with a detailed record of the early universe. As we will see in Chapter 6, various anomalies have been reported in the CMB at long wavelengths or large distances— which is where we expect to find traces of what the universe was like before quantum death took over. Remarkably, the reported anomalies are of the kind we would expect to see from an early breakdown of Born's formula. It is, however, difficult to make accurate measurements of the CMB at large distances, so the evidence is hard to evaluate precisely. This is still a topic of current research and as yet there are no definite answers—only intriguing hints.

The reader might be wondering why this is all so fiendishly difficult. Thinking back to the feared heat death of the universe—well, just look around us, here we are 13.8 billion years after the Big Bang and the heat death has not happened. The Sun and other stars are still shining, with billions of years to go before burning out. If heat death has not happened yet, why should quantum death have happened already? This is a good question.

We can put the question another way. In today's world we do not see the same thermal noise everywhere—and yet we *do* see the same quantum noise everywhere. Why the difference?

This raises some fascinating questions about gravitation. The reason we are not already overwhelmed by heat death is that gravity causes matter to clump together. The early universe was almost completely smooth and the temperature was almost completely uniform. The heat death had almost been reached. But some regions were a little bit denser, hence they had a slightly bigger gravitational pull on their surroundings. This means they pulled in more matter and became even denser. This process is called 'gravitational clumping'. It explains where stars and galaxies ultimately came from. Our universe started out looking much the same everywhere (smooth and homogeneous, like a gas or liquid). Because of gravitational clumping, over time some regions became more and more dense compared with the rest. As the material grew denser it became hotter, eventually igniting thermonuclear reactions. We now call these clumped regions 'stars'. They are the fires we see in the sky, including our Sun. This is ultimately why we are not trapped in a heat death today. We have been saved, as it were, by gravity. This raises the tantalising possibility: could gravity somehow save us from quantum death as well?

This brings us to our third loophole. To understand this we will have to rethink some of the mysteries of 'quantum gravity' and of black holes (lively and contentious topics of current research). As we will see in Chapters 8 and 9, the difficulties we face at the deepest level of gravitational physics might be solved if we accept that gravity can counteract quantum death and free us from the debilitating effects of quantum fog. According to these ideas, the precious resource of subquantum matter can be *created* by gravity—in particular by exploding black holes. We could, in principle, collect such matter and put it to technological use. We would then have well and truly escaped from quantum death.[10]

---

[10] A fourth loophole, not explored in this book, considers the breakdown of de Broglie's law of motion at 'nodes' where the pilot wave has zero magnitude and the

It is, however, a long road to quantum reality and quantum life. We now move on to a closer look at pilot-wave theory: how it works, how it was discovered, and some of the many misunderstandings that still surround it.

'phase' of the wave is undefined. We have argued that new physics must set in sufficiently close to such points, and this might provide another way to counteract quantum death. Such effects could show up, for example, in high-energy particle collisions. However, these ideas are still being developed.

2
Seeing the Impossible

Quantum pornography

In 1985, at a scientific conference in Italy, a young British physicist by the name of Chris Dewdney screened a film showing a particle moving towards a hard barrier. The lights were dimmed as the old 16mm projector whirred into action. The audience gasped as the particle approached the barrier, seemed to pause and hesitate, and then made its way through to the other side with a curious wriggling motion. According to the classical physics of Newton and Einstein, this was impossible. The particle should have hit the barrier and bounced right back. The process of 'quantum tunnelling'—which allows a particle to break through an impassable barrier, and which occurs in radioactive decay—was thought to be impossible to explain or visualise in terms of moving particles. And yet here it was for all to see.

There was more. Dewdney showed his audience pictures of a particle moving towards a barrier with two holes (or slits) cut into it. The particle was seen to approach the barrier and go through one of the holes—emerging on the other side, again, with a curious wriggling motion. The particle eventually landed on a 'backstop' (Figure 25a). Dewdney explained that, if this was repeated for a different particle starting at a different point, the particle would again go through one of the holes (perhaps

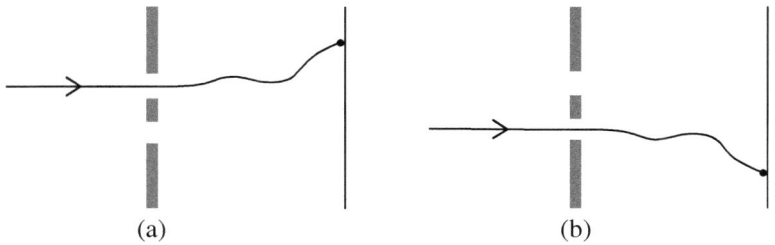

Figure 25. Images of a moving quantum particle as shown by Chris Dewdney in 1985.

Beyond the Quantum. Antony Valentini, Oxford University Press. © Antony Valentini (2025).
DOI: 10.1093/oso/9780198853749.003.0003

the other one this time) and now land somewhere else on the backstop (Figure 25b). This process could be repeated, again and again, with different particles starting at different points. A pattern of dots or blips builds up at the backstop showing where the particles land (Figure 26a). The particles are seen to cluster in some regions and avoid others. This can be represented by a wavy pattern: many particles land where the wave is large, few particles land where the wave is small (Figure 26b).[11] The wavy pattern matches the famous 'quantum interference' seen in laboratory experiments. Dewdney claimed that, in these images, we were witnessing the real behaviour of quantum particles.

Many in the audience were stunned. What they were seeing was widely regarded as impossible. Some of the greatest physicists of the twentieth century had assured them that, in the 'two-slit experiment', a single quantum particle could not travel through only one hole. Instead, each particle somehow had to go through both holes at once. And yet here we could *see* quantum particles going through just one hole at a time. How?

When Dewdney finished speaking, the Italian physicist Franco Selleri stood up in excitement and exclaimed:

This is the hardcore pornography of quantum physics!

According to Selleri, this sort of thing had been 'censored' for decades.

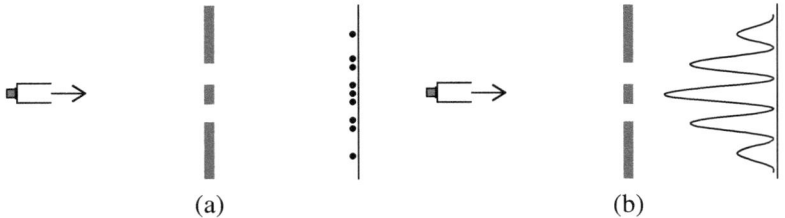

Figure 26. Quantum interference as shown by Chris Dewdney in 1985. One particle at a time is fired at a barrier with two holes and lands somewhere on the backstop. After many repetitions the clustering in some regions (part (a)) can be represented by a wavy pattern (part (b)).

[11] Technically, the number of particles landing close to a point is given by the square of the height of the wave at that point.

Dewdney's talk had made a deep impression. It was a remarkable achievement for a young unknown attending his first major international conference. But what exactly had Dewdney done—and why was it so astounding?

The 'only mystery' of quantum mechanics

The two-slit experiment is widely hailed as the most iconic experiment in quantum physics—*the* experiment that delivers a killer blow to our naïve and preconceived ideas about reality. Here is how it works.

We fire a particle at a barrier with two holes in it (Figure 27). The particle might be an electron (the smallest atomic particle), or perhaps an atom or a molecule. In any case we fire the particle at the barrier and record where it lands on the other side (as shown by a blip somewhere on the backstop). It should be emphasised that we do not watch the particle as it leaves the source and travels through the barrier, we only look to see where it lands. Now we can repeat this again with another similar particle. And then another, and another. A wavy pattern of blips forms at the backstop. This is often called an 'interference pattern'. The experiment has been performed many times, in countless laboratories all over the world. There can be no doubt about the result. But what does it mean?

The textbooks describe the experiment as follows. There is no particle moving towards the barrier. But there is a quantum wave. When the wave hits the barrier it passes through the slits, so that two waves emerge

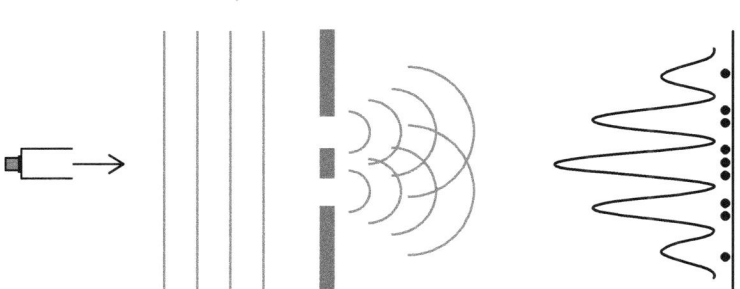

Figure 27.   The iconic two-slit experiment. A particle fired at the barrier (with two holes) lands somewhere at the backstop. With many repetitions the particles land at different points, clustering in a wavy interference pattern. According to quantum mechanics, the particles do *not* move from source to backstop: they simply appear when observed.

on the other side (Figure 27). The waves overlap forming an interference pattern—think of dropping two stones into a pond and watching the ripples overlap. So the wave comes in through the two slits and forms a wavy pattern at the backstop. As for the particle, we do *not* imagine it actually moves from the source to the backstop. It starts at the source and somehow ends up at the backstop (when observed), with nothing definite in between. Nor can we predict where the particle will land. But we can predict the probability for where it will land. Born's formula tells us that, if we repeat the experiment many times, the fraction of particles landing near a certain point is given by the squared magnitude of the wave at that point. And so, after many repetitions, the blips recorded at the backstop form the famous wavy pattern shown in Figure 27. That is what quantum mechanics tells us.

The wavy pattern is intriguing. Surely, you might think, there must be an explanation for it in terms of moving particles. Well, not according to received opinion. The Nobel Prize-winning American physicist Richard Feynman famously said that the wavy pattern contains 'the *only* mystery' of quantum mechanics—by which he meant that all the mysteries of quantum mechanics are in a sense reflections of this one basic mystery.

Despite the warnings, let us give in to temptation and try to come up with an explanation in terms of moving particles—if only to learn why it cannot be done. To explain the wavy pattern, perhaps we need to understand exactly how the particles move around. We might suppose that each particle emerges from the source, moves towards the barrier, passes through one of the holes, and eventually lands at the backstop, as in Dewdney's 'pornographic' pictures (Figure 25). Clearly, the particles will have to move in strange ways for them to end up clustering with the wavy pattern seen at the backstop. According to received opinion, however, such an explanation is not merely difficult but *impossible*. Here, for example, is Feynman commenting on this idea, in his famous physics lectures delivered to undergraduate students at the California Institute of Technology in 1965:

> Many ideas have been concocted to try to explain [the wavy pattern] in terms of individual electrons going around in complicated ways through the holes. None of them has succeeded.

Feynman made it sound as if there had been a long history of complicated and failed attempts to explain the wavy pattern in terms of moving particles. He also claimed that any such explanation was doomed to fail.

According to Feynman (and many others before and since), if each particle did in fact get from source to backstop by passing through one of the holes, we would never see the wavy pattern. Since we do see the wavy pattern, we have experimental proof that the particles do not go through one hole or the other. As Feynman put it:

It is *not* true that the electrons go *either* through hole 1 or hole 2.

That sounds like some pretty solid science—which a Nobel Prize-winning physicist taught to undergraduates at one of the world's leading research universities. So let us examine the argument. We are going to assume that each particle really does go through one hole or the other, and from this we are going to show that there can be no wavy pattern (in contradiction with experiment).

Here we go. If we assume that each particle goes through one hole or the other, then, obviously, some of the particles go through the top hole, and the others go through the bottom hole. Let us now focus our attention on the particles that go through the top hole. It might be thought that the motion of those particles must be unaffected if we close the bottom hole. So we can find out how those particular particles cluster at the backstop by doing the experiment with the bottom hole closed. The results of that experiment are shown in Figure 28a. As we might expect, the particles simply cluster in front of the top hole. There is now no wavy pattern, just a smooth hump showing where the particles land.[12]

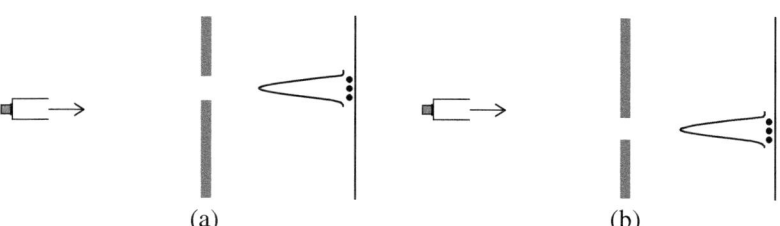

Figure 28. Two-slit experiment with (a) the bottom hole closed, and (b) the top hole closed.

[12] We are simplifying slightly. The curve shows a bit of wave-like 'diffraction' at the edges. We can ignore this here and take the curve to be a simple smooth hump.

Returning to the original experiment with both holes open, the same reasoning applies to the particles that go through the bottom hole. Their motion must be unaffected if we close the top hole. So we can find out how those particular particles cluster at the backstop by doing the experiment with the top hole closed. The results are shown in Figure 28b. The particles now cluster in front of the bottom hole, again with no wavy pattern and just a smooth hump.

Let us now apply what we have learned to the original experiment with both holes open. We are assuming that each particle goes through either the top hole or the bottom hole. We have argued that those that go through the top hole will cluster at the backstop as shown in Figure 28a, while those that go through the bottom hole will cluster at the backstop as shown in Figure 28b. It follows that, with both holes open, and now considering all the particles together (regardless of which hole they go through), the clustering at the backstop must look like the double hump shown in Figure 29— obtained by combining the two curves from Figure 28a and 28b.

And now—*drum roll*—the crux of the killer argument. At this point the reader might be excited: we are about to find out exactly why naïve ideas about reality fail in quantum physics! Here it comes. We have deduced that, with both holes open, the clustering at the backstop must look like the double hump shown in Figure 29. But the conclusion is wrong. When we carry out the experiment with both holes open we do *not* see a double hump. Instead we see a wavy pattern as shown in Figure 27. So our argument

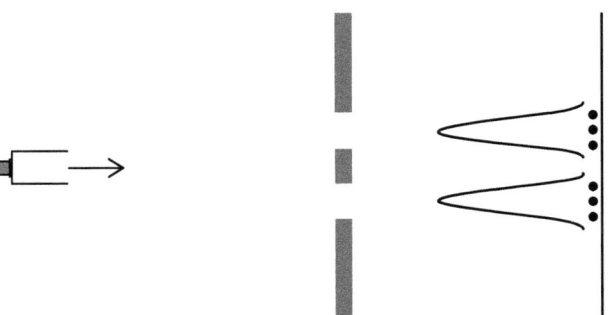

Figure 29.   Killer argument against quantum reality? If each particle really goes through one hole or the other, then with both holes open we (supposedly) *must* find a simple double-hump clustering at the backstop, contradicting the wavy pattern we see in the lab.

contradicts experiment. And so something is wrong with our argument. Logically, we must have made a wrong assumption somewhere along the way. But where, exactly, does the blame lie?

For most of the twentieth century physicists like Feynman blamed the assumption that each particle goes through either the top hole or the bottom hole. This can seem so innocuous as to hardly be an assumption at all. And yet physicists usually put the blame there: our picture of particles moving in definite ways, through one hole or the other, is just plain wrong. And the two-slit experiment *proves* it to be wrong. If each particle really did go through one hole or the other, the clustering at the backstop would be the double hump shown in Figure 29—and not the wavy pattern we see in Figure 27 and in the lab. Therefore we can safely conclude that each particle does not go through one hole or the other. End of killer argument.

It seems that our attempt to understand the wavy pattern in terms of particle motion has failed. We were naïve to assume that each particle goes through one hole or the other. Perhaps each particle goes through both, or neither. Maybe we should not think about it. Just shut up and calculate the wavy pattern—and accept it as one of those unfathomable mysteries.

Except for one thing: what about Dewdney's pictures?

A physics of the invisible

We can now understand why Dewdney's conference presentation caused such a stir in 1985: his pictures flatly contradicted basic physics. It was not possible for a quantum particle to go through one hole or the other. Every undergraduate physics student knew that. And yet, there it was for all to see. How had Dewdney done it?

We should clarify right away that Dewdney's pictures (as well as his cine film) were not made by photographing (or filming) particles moving in real laboratory experiments. As we have said, in a two-slit experiment we do not watch the particles as they emerge from the source, we only record where they land. Dewdney's pictures were made, not by recording real experiments, but by *calculating* how the particles move, according to a theory that had long been forgotten. In effect, Dewdney had a way of working out, mathematically, what each particle is doing when no one is looking at it. But how?

Dewdney had performed his calculations some years earlier, in 1978, when he was a PhD student at Birkbeck College in London. Dewdney was working in a research group centred around David Bohm. The point was to illustrate a theory of particle motion that Bohm had developed in the early 1950s. As we saw in Chapter 1, Bohm's work was really a revival of the pilot-wave theory proposed in the 1920s by Louis de Broglie. In this theory a moving particle is guided by a pilot wave (as a floating bottle is carried along by an ocean wave). Dewdney had applied the mathematical laws of pilot-wave theory to calculate how particles move through a two-slit experiment—when no one is looking at them. This is much the same as applying Newton's laws to calculate the motion of the Moon when no one is looking at it.

Referring to Figure 30, the pilot wave comes in from far away and carries the particle along in a straight line. When the wave hits the barrier, two waves emerge on the other side, one from each hole (as would happen for an ocean wave hitting a hard barrier with two holes in it). On the far side of the barrier, the two waves combine and jostle the particle giving it a wriggling motion. Simple enough.

The exact motion of the particle can be calculated from de Broglie's law of motion. The mathematical details will not be needed here, but roughly the law is this: the particle moves perpendicular to the crests of the wave, moving faster where the wavelength is shorter and slower where

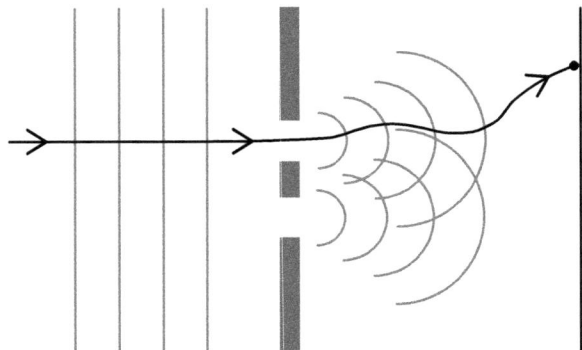

Figure 30. The 'hardcore pornography' of quantum physics. Particle motion in a two-slit experiment, as first calculated by Chris Dewdney using the equations of pilot-wave theory. Each particle is carried along by the pilot wave and travels through one hole only.

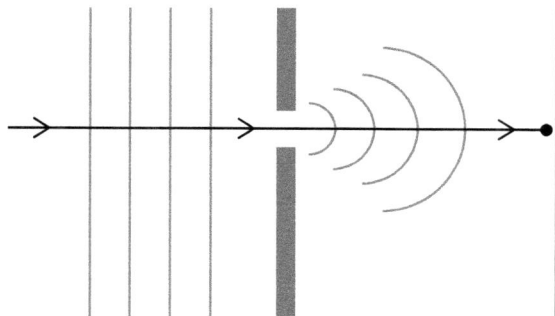

Figure 31. Killing the killer argument. Closing one hole changes the pilot wave and so changes the motion of the particle through the other hole (compare Figure 30).

the wavelength is longer.[13] When applied to a two-slit experiment, as done by Dewdney, we find a strange wriggling motion where the two waves overlap and interfere (Figure 30).

Now we can see what went wrong with Feynman's argument. Remember we also assumed that a particle moving through one hole is not affected by closing the other hole. After all, if the particle is moving through one hole how can it possibly 'know' if the other hole is closed or not? And yet, in pilot-wave theory, that assumption is wrong. Closing one of the holes *does* affect the particle moving through the other hole. And for a simple reason: closing one of the holes changes the pilot wave acting on the particle, and so changes the motion of the particle (see Figure 31, and compare with Figure 30). The same thing would happen with an ocean wave: closing one of the holes would change the wave carrying a floating bottle on the far side, and so change the motion of the bottle. This is how pilot-wave theory can explain the wavy pattern, with each particle moving through one hole or the other—despite a century of claims that this is impossible.

Let us be clear. The argument predicting the double-hump clustering of Figure 29 must have gone wrong somewhere. Physicists usually blame the assumption that each particle goes through one hole or the other. But another assumption was made along the way: that closing one hole does not affect the particle moving through the other hole. According to pilot-wave theory, *there* lies the mistake.

---

[13] Technically, the particle velocity is proportional to the phase gradient.

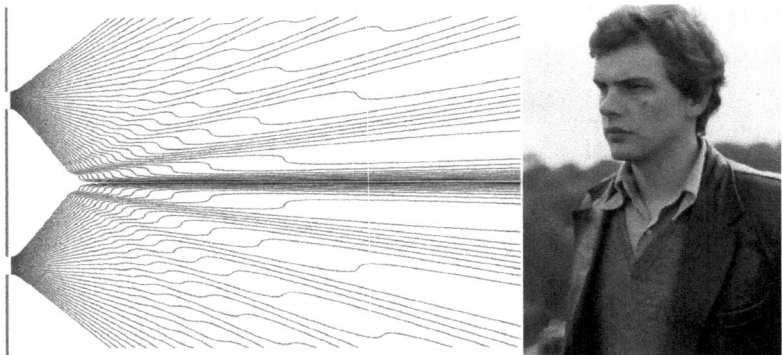

Figure 32.   Seeing the impossible. Particle trajectories in a two-slit experiment (left) as calculated in 1978 by British PhD student Chris Dewdney (right).

Once this is understood, it is easy to explain the two-slit experiment. Dewdney calculated what happens as the incoming particles approach the barrier from different points. If we fire lots of particles at the barrier, one at a time, we will find lots of wriggling trajectories landing at different points on the backstop. Dewdney's results were first published in 1979 and are shown in Figure 32.

In a sense what Dewdney had done was not really new. It had already been understood by de Broglie in the 1920s, and again by Bohm in the 1950s. But no one had worked out the details of the particle trajectories. Dewdney—a modest, quietly spoken man with piercing sky-blue eyes— was passionate about *seeing* what the trajectories looked like. At that time virtually every physicist in the world believed that it was simply not possible to draw realistic pictures of quantum processes. To see such pictures emerging from his computer simulations must have been an exhilarating experience for a PhD student living on a shoestring in 1970s London. In later years, Dewdney went on to explore other quintessentially quantum phenomena, showing precisely and in detail how pilot-wave theory could explain them in terms of moving particles.

Dewdney's early work on the two-slit experiment had a startling effect, at least on those who were receptive to it. Remarkably, by the 1970s, Bohm had lost interest in what we now call de Broglie–Bohm pilot-wave theory. But Dewdney's calculations rekindled his interest. Together with students and other collaborators, Bohm began working on the theory once again, publishing papers that started to attract attention, culminating in a book (co-authored with British physicist Basil Hiley) published in 1993. In the same year, British physicist Peter Holland—at that time working in Paris—published what amounted to an undergraduate textbook on quantum mechanics written in terms of pilot-wave theory. It was becoming increasingly difficult for the physics establishment to ignore this theory, although misunderstandings remained widespread.

Also influential were the writings of John Bell, a Northern-Irish particle physicist working at the European Organisation for Nuclear Research (CERN) in Geneva. Bell had done significant work in particle physics, but was more widely known for his breakthrough paper of 1964 in which he showed that any realistic explanation for quantum entanglement would have to involve nonlocality. Specifically, 'Bell's theorem' shows that a realistic explanation for quantum statistics requires entangled particles somehow to communicate instantaneously across space.[14] Bell was one of the very few mainstream physicists supporting pilot-wave theory, in part because he had concluded from his theorem that nonlocality was not merely a peculiarity of pilot-wave theory but a feature of nature. In 1987 Bell published a widely acclaimed book that was highly critical of quantum mechanics. The book included several chapters explaining how pilot-wave theory worked, much to the bemusement of many of his colleagues who had been trained to believe that such theories were impossible. By the early 1990s, a small but growing number of researchers were re-examining this old theory from the 1920s—a theory which was still widely dismissed as impossible but which, plainly, was perfectly possible.

What about quantum randomness?

When we fire a particle at a barrier with two slits, we see that it lands somewhere on the backstop. If we fire another particle, we see that it lands somewhere else on the backstop. Again and again, firing one particle after another, they generally land at different places. Why? After all, we are doing the same experiment with each particle. Why do they not all land in the same place?

---

[14] With some caveats: see the final section of Chapter 3.

Physicists tell us that that we are witnessing the fundamental randomness of nature. In quantum physics, 'God plays dice'. No reason can be given for why this particle landed here and that particle landed over there. All we can talk about are probabilities and statistics—how many particles are *likely* to land here or there in a given experiment. For a large number of particles, we can calculate the percentage that will land in a certain region. We do this using Born's formula: the probability of landing somewhere is given by the squared magnitude of the quantum wave. Nothing more can be said. To ask *why* a specific particle lands at a specific point is to dabble in meaningless metaphysics.

Perhaps we can find a better answer. Let us think for a moment, not about quantum particles, but about tennis balls. When you throw a tennis ball, why does it not always land in the same place? Even without any formal training in physics, you already know the answer. Maybe last time you threw it a bit harder so it went further away, while this time you threw it a bit softer and it landed closer to you (Figure 33). Maybe you are not throwing it in exactly the same direction each time, so again where it lands changes a bit from one throw to the next. And finally, maybe you are not standing in exactly the same place each time you throw the ball—that will also affect where it lands. So where the ball lands depends on three things: its speed, direction of travel, and position (at the time it leaves your hand).[15] These factors are often called 'initial conditions'—they determine what happens later. We then have a simple explanation for the different outcomes: the tennis ball does not always land in the same place because the initial conditions vary from one throw to the next.

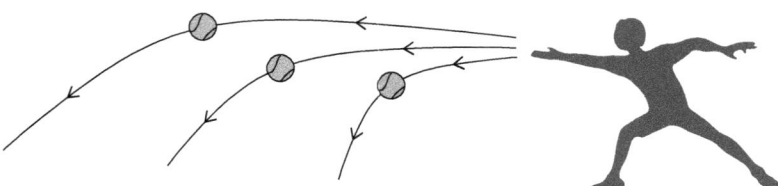

Figure 33.   Why does a thrown tennis ball not always land in the same place? Because the initial conditions vary from one throw to the next.

[15] A gust of wind will also affect where the ball lands, but to keep things simple we can ignore that and assume there is no wind.

Now let us think again about firing a quantum particle at a barrier with two slits. After each shot, why does the particle not always land in the same place on the backstop? We have been told that the scatter of outcomes is a sign of 'God playing dice with the universe'. Well, perhaps. Or could it just be that the initial conditions are a bit different, from one shot to the next, as in the case of throwing a tennis ball? Maybe we are not firing the particle in exactly the same way each time, and *that* is why it lands at different places? According to pilot-wave theory, that is in fact the right answer. The particles land at different points on the backstop, not because of any fundamental randomness, but because they *start* at different points as they emerge from the source. Where the particle lands depends on where it starts—simple as that.[16]

We can now dig more deeply into the 'only mystery' of quantum mechanics. We have said that the particles land at different points because they start at different points. It is important to be clear about this. If instead each and every particle started at the same point, then each and every particle would land at the same point. Instead of a wavy pattern we would see only a single blip or small blob, with all the particles clustered at the same point on the backstop (as in Figure 30). So our original question—why do the particles not always land at the same point?—has now been pushed back to an earlier question: why do the particles not always start at the same point when they emerge from the source?

In the case of tennis balls, as described by classical Newtonian physics, there is no reason why they could not be thrown from the same point (and with the same speed and direction of travel). They would then all land at the same point. We might find this difficult to do in practice, but in principle it could be done accurately, and (with enough effort) as accurately as we like. How about doing the same with quantum particles? Come on, you might say, let us really make an effort and perform a two-slit experiment more carefully. If we fire each and every particle from the same starting position, then each and every particle will land at the same point on the backstop. There will be no wavy pattern, and that will be the end of this quantum nonsense! That may sound straightforward, but it is currently impossible.

---

[16] In pilot-wave theory the only thing that varies from one shot to the next is the initial position. The other conditions—the initial speed and direction of travel—are fixed by the quantum wave.

There is always some scatter in the starting positions. So there is always some scatter in the outcomes—like it or not.

Now we have really hit bedrock. There is no quantum randomness. Each outcome is determined by the starting conditions. But the starting conditions vary from one particle to another—and *that* is the mystery we need to focus on.

It is as if something always prevents us from 'throwing the tennis ball' in exactly the same way. But what? According to quantum mechanics, the scatter in the starting positions is inevitable: it is dictated by the uncertainty principle, which most scientists believe is a rock-solid law of physics. The bedrock cannot be broken. But pilot-wave theory says *no*. The uncertainty principle is not an unbreakable law of physics—it applies only to the special state of quantum death which we happen to be confined to.

As we saw in Chapter 1, in the state of quantum death everything is permeated by quantum noise—everything, including particles emerging from our source in a two-slit experiment. The state of quantum death was probably reached in the early universe soon after the Big Bang. And so, today, we live in a world where all quantum particles have an irreducible uncertainty or scatter in their positions. There are possible ways out of this—involving relic particles from the early universe, or exotic gravitational effects—but right now we are stuck with what we have. Everything we put a hand to is tainted by quantum noise—and *that* is the ultimate reason why we cannot fire different particles at a two-slit barrier with the same starting positions.

We should point out that in 1978 Dewdney did not concern himself with the origin of quantum noise. Like most workers in the field he simply assumed that the starting positions varied from one particle to the next, so that the outcomes would vary from one particle to the next, in such a way that we obtain the wavy pattern at the backstop. This brings us to another important point.

Born's formula and quantum death

When we look at the scatter of outcomes in a two-slit experiment, we do not see any old pattern: we always find the same wavy pattern as predicted by Born's formula (Figure 27). To get the right pattern at the backstop, Dewdney not only had to assume a scatter in the starting positions—he had to assume a certain *kind* of scatter. Specifically, he had to assume that the starting positions were already scattered according to Born's formula. In other words: he had to assume the right quantum noise coming in from the source, in order to get the right quantum noise coming out at the backstop. What comes in comes out, you might say.

De Broglie's law of motion shows that this is always true: if we assume that the initial conditions are scattered according to Born's formula, then inevitably the final outcomes will also be scattered according to the same formula. This is the state of quantum equilibrium: once we have Born's formula, we stay with it.

Dewdney assumed that the particles coming in from the source already follow Born's formula. Neither he nor his colleagues asked why. But we can ask why. The particles coming in from the source have a scatter in their initial positions, but why does the scatter always match Born's formula? The answer has already been given, in outline, in Chapter 1: the starting positions are scattered according to Born's formula because the particles have a past history during which they experienced quantum relaxation. Over time the peculiar wriggling motion of quantum particles leads to the end state of quantum equilibrium, in which the particle positions are distributed according to Born's formula. And once we get there, we stay there.

We call the end state 'quantum death' because of the serious limitations it imposes on us. Recall the analogy between quantum death and the heat death of the universe (Chapter 1). If everything in the universe had already reached the same temperature, then everything would be jittering in the same way (with the same thermal or Brownian motion). And there would be no escape. Similarly, once the fog of quantum noise has permeated the universe, there is no escape—not even in the laboratory.

So the particles coming in from the source obey Born's formula because they have already relaxed to the equilibrium state. And they probably did so a very long time ago, near the beginning of the universe. But, one day, we might discover exotic particles that do not obey Born's formula. If we fired such 'nonequilibrium' particles at a two-slit barrier, we would not see the usual wavy pattern at the backstop. Instead, we would see something else. The pattern could be blurred, or parts of it might be missing (Figure 34). Quantum physics would be broken. According to pilot-wave theory, this is quite possible. Once we leave the confines of quantum death, there is a whole new world of radical physics to explore.

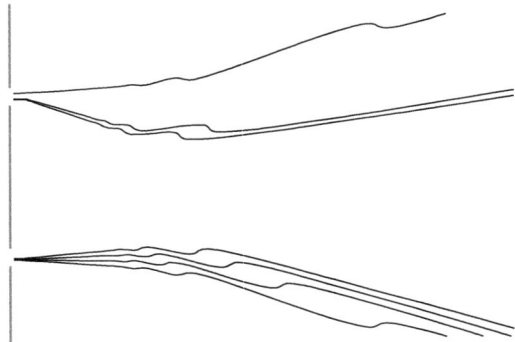

Figure 34. Breaking Born's formula in a two-slit experiment. If the particles fired at the barrier do not have the conventional spread of positions predicted by quantum mechanics, what comes out on the other side can look quite different from what is usually observed (compare Figure 32).

But we are getting ahead of ourselves. These are subjects for later (Chapter 5). Dewdney just assumed that Born's formula was true for the initial conditions and left it at that. So did de Broglie, and so have most workers since. We then always find the same wavy pattern as predicted by quantum mechanics.

Explaining nonlocality

We said in Chapter 1 that entangled particles seem to be directly and instantaneously connected with each other over large distances. This mysterious connection is called nonlocality. How does pilot-wave theory explain it?

In classical physics, in particular as elaborated by Einstein, we think of the world as made of separate objects or parts in space—like pieces of a jigsaw puzzle that fit together. Each piece is directly connected only with neighbouring pieces, not with pieces far away. This is called 'locality' (Figure 35).

You might say: wait a minute, I can stand here and call out to my friend over there, who hears me even when they are quite far away. How does that work? It is still local action. Your vocal cords make the air in your larynx vibrate. The air molecules knock against each other (when they are nearby), creating regions of high and low pressure that push against each other and move rapidly outwards—what we call a sound wave. It travels at a speed of about one fifth of a mile per second. When the sound wave

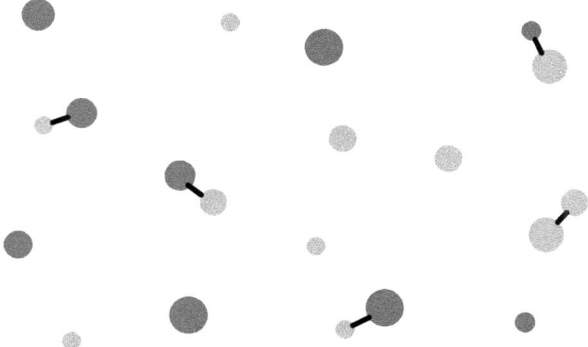

Figure 35. Classical physics and locality. The world is made of parts that affect each other (black connecting lines) only when they are nearby.

reaches your friend's ear, the oscillating pressure makes their eardrum vibrate (Figure 36). So you have made your friend's eardrum vibrate, not directly but *indirectly*—through the local action of an intervening medium (the air).

This example, simple as it may be, conveys the essence of much of classical physics. Consider, for example, 'electromagnetic forces'. A charged particle here does not act directly on a charged particle over there. Instead it acts indirectly via an intervening medium—what we call the 'electromagnetic field'. If we shake a charge here, it causes a ripple in the electromagnetic field *here*, which spreads outwards at the speed of light (about 186,000 miles per second) and which eventually hits another charge and knocks it around. That is how radio works: a transmitter shakes charges up and down to create radio waves that eventually reach a distant receiver and shake charges up and down inside it. This is pretty much like dropping a stone into a pond: the water waves ripple outwards, and when they hit a faraway bottle floating in the water they make the bottle bob up and down. This is all interesting physics—caused ultimately by local action, that is, by one thing knocking against something else close to it, which then knocks against something else close to *it*, and so on and on (like a domino effect).

In classical physics, then, widely separated objects can be connected, but only indirectly by means of an intervening medium or field.[17] And there

---

[17] This is true in classical physics as modified by Einstein. In Newton's original theory of gravity, there is action at a distance without an intervening medium. However, its

Figure 36. Local action. A pressure pulse (or sound wave) created when you speak travels through the air and reaches your friend's ear.

is one more thing. When the local action spreads out from one place to another, there is an upper limit on how fast it can go. According to Einstein's relativity, no physical influence can move faster than the speed of light. Water waves are very slow, sound waves are faster, but light waves (or electromagnetic waves) are the fastest of all. Nothing—we repeat, nothing—can travel faster than light.

In quantum physics, in contrast, it seems that widely separated objects can be connected *directly*, with no intervening medium or field. And they can affect each other *instantaneously*, no matter how far apart they may be (Figure 37).

To those with little background in physics, this may not seem so shocking. But most physicists *are* shocked. To convey a sense of why, imagine this. Your friend who became an astronaut is out in space, far above the Earth. She does not have a radio or a phone or any other device that can shake the electromagnetic field. She is wearing a space helmet. Outside the helmet there is no air, just the vacuum of space (Figure 38). You are at home sitting in an armchair. She speaks—*and you hear what she says*, instantly and crystal clearly, as if she were sitting right next to you. Would you not be surprised?

Famously, in a letter to Born written in 1947, Einstein dismissed nonlocality as 'spooky'. And yet, in recent decades, experiment after experiment has verified its existence (up to some caveats).[18] With reasonable assumptions, the direct action at a distance that so disturbed Einstein is real. Even so, it turns out that we are unable to *make use of* nonlocality to send messages. We can deduce that nonlocality is there, but we cannot control it. So

effect decreases rapidly with distance (as the inverse square), and so it is not really comparable to the nonlocal action at a distance found in pilot-wave theory, the effects of which do not depend on distance at all.

[18] For a full discussion see, again, the final section of Chapter 3.

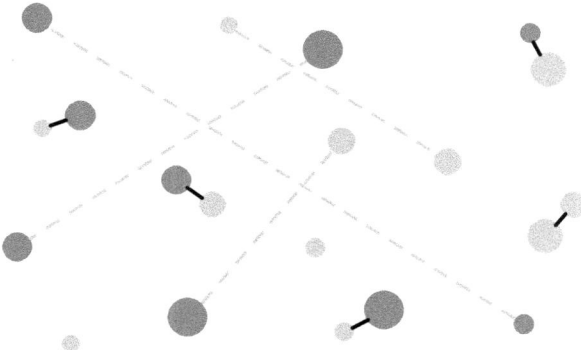

Figure 37. Quantum physics and nonlocality. The world is made of parts that can affect each other directly and instantaneously (dashed lines)—even when they are widely separated and there is no intervening medium or field.

while most physicists still squirm in their seats when they think about non-locality, there is no overt conflict with Einstein's relativity—the conflict is there, but hidden.

The American physicist and philosopher Abner Shimony was fond of saying that in modern physics there is a 'peaceful coexistence' between relativity and quantum mechanics. The two theories are opposed to each other, but there is no actual warfare. Remote objects are directly and instantaneously connected, but we cannot make use of those connections for practical signalling. Nonlocality is true in principle, but in practice relativity is still correct.

According to pilot-wave theory, the peaceful coexistence Shimony talked about is a peculiarity of quantum death—the special state we are confined to, where quantum noise pervades all things. It is only in this state that we are unable to control nonlocality. We are then unable to send instantaneous messages and Einstein's relativity remains unthreatened (at least in practice). If instead we could escape quantum death, we *would* be able to send instantaneous messages—from inside an astronaut's space helmet far above the Earth, directly into your living room. Einstein's relativity would fall apart. Before discussing that, we need to understand nonlocality better.

How exactly does pilot-wave theory explain nonlocality or instantaneous action? The essence of the answer is that, at the deepest level of pilot-wave theory, the world is not really made of separate parts located in space.

Figure 38. Shocked by nonlocality. There is no air in space to transmit sound. The astronaut has no radio. And yet, when she speaks, you hear it instantly. How?

What we see as widely separated objects are really different components of one object in a higher-dimensional 'configuration space'—the space that defines the state of everything in the universe. The pilot wave acts in this higher-dimensional space. When we look at one piece of the universe, it seems to be mysteriously connected with other pieces that are far away in space, but that is because they are really all one object. This may sound outlandish—but that is what the theory says.

To get an intuitive sense of what this is about, consider a crude analogy. Imagine two people looking at the same tennis ball from different angles. As the ball moves through the air, what one person sees will be different from what the other person sees—for example, one might see it moving further away while the other sees it moving closer. Perhaps they do not realise they are looking at one and the same ball. But then something interesting happens: whenever the ball hits the ground, they *both* see it hit the ground—and at the same time. And so, comparing notes, they begin to suspect they are looking at one and the same ball. This analogy misses a lot, but it is a start.

To see what is happening more accurately, let us think about two entangled particles guided by de Broglie's pilot wave. The particles move in our everyday three-dimensional space (Figure 39a). That means we need three numbers to specify where each particle is. To specify where both particles are we then need six numbers. And so we have a six-dimensional configuration space. The pilot wave acts on the system in this six-dimensional

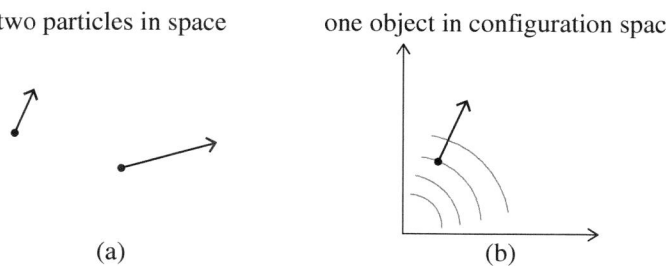

Figure 39. What we see as two particles moving in space are really one object moving in a higher-dimensional configuration space (guided by a pilot wave).

space. We cannot draw that, but Figure 39b gives a simplified picture in two dimensions. While Figure 39a shows two particles moving in ordinary space, Figure 39b shows one object moving in a higher-dimensional configuration space. The wave crests represent the pilot wave. The object is carried around by the wave, and its motion represents the motion of *both* particles.

It might seem that Figure 39a and 39b are just mathematically equivalent pictures—two different ways of describing the same physics. But according to pilot-wave theory, the second picture is closer to reality. And here is why. Let us say an experimenter does something to push one of the particles—perhaps they knock it with a stick. We might think this would affect only that one particle, but we would be wrong. In pilot-wave theory, knocking one particle with a stick has the extraordinary effect of changing the whole pilot wave. This then changes the motion of the *other* particle as well.[19] In other words: if we hit this particle here with a stick, the other particle instantaneously 'feels it'.

If you are finding this difficult to visualise, you are in good company. But the mathematics can be worked out, even if we cannot visualise it very well. In a real experiment, in a real physics lab, we probably would not try hitting a tiny electron with a stick. But we might try deflecting it with an electric field. An electron carries a charge, so when it enters an electric field there is a force pushing on it. So now imagine we have two entangled electrons, which might be far apart. Let us take this one here and switch on

---

[19] Technically, if we change the 'local Hamiltonian' for one particle this will instantly affect the velocity of the other particle (if the particles are entangled).

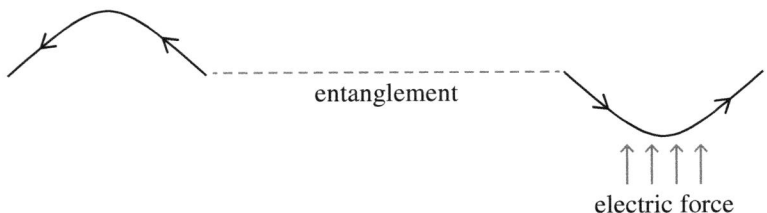

Figure 40. Nonlocal action. Deflecting a particle here also deflects a particle over there—instantaneously, no matter how far away.

an electric field to deflect it (Figure 40). According to pilot-wave theory, the other electron is also deflected. In other words: deflecting a particle here somehow also deflects a particle over there. It is as if you hit your tennis ball with a racquet, and someone *else's* tennis ball also goes flying through the air—the two balls are 'one'.

3
Origins of Pilot-Wave Theory

Louis de Broglie in 1920s Paris

Where had pilot-wave theory come from? To understand its origins, we must backtrack from an impoverished PhD student in 1970s London to an independently wealthy PhD student in 1920s Paris.

We have already briefly met Louis de Broglie, the founder of pilot-wave theory. He was in fact a scion of the French aristocracy, born as Prince Louis Victor Pierre Raymond de Broglie, from a distinguished line that included dukes, princes, ambassadors, and marshals of France (originating from Italy in the seventeenth century, with the family name Broglia). By all accounts modest and solitary, with the air of an outsider, de Broglie was deeply read in history. Like his older brother Maurice (the sixth duc de Broglie), who became a distinguished experimental physicist, Louis resisted family pressures to serve France and instead devoted himself to science. Both brothers might have spent their time and resources on the traditional aristocratic pursuits of horses and game hunting. Instead, they devoted themselves to hunting the biggest game of all: an understanding of reality at the deepest level of science.

Maurice had set up his own private laboratory in the family mansion in Paris, on the rue Châteaubriand. He carried out ground-breaking work with X-rays, deepening our understanding of how they interact with matter. At that time almost all physicists believed that X-rays were simply waves—electromagnetic waves, just like light or radio waves, albeit of much shorter wavelength. Since 1905 Einstein had been virtually alone in his belief that electromagnetic waves contained particle-like concentrations of energy—what we now call 'photons'. In other words, Einstein believed that X-rays, light, and other electromagnetic waves, were all made of both particles and waves. In the early 1920s, during conversations with his brother, Louis de Broglie became convinced that Einstein was right.

Beyond the Quantum. Antony Valentini, Oxford University Press. © Antony Valentini (2025). DOI: 10.1093/oso/9780198853749.003.0004

Then, in the summer of 1923, de Broglie had an extraordinary idea: that ordinary matter—the solid matter we see around us—was also made of both particles and waves. An atomic particle such as an electron was not simply a particle moving around in space, it was also accompanied by a guiding wave or 'pilot wave'.

De Broglie's idea had several motivations. He was intrigued by a well-known mathematical analogy between the theory of optics and the theory of mechanics. This long-standing analogy had never been explained. If particles were in fact guided by waves, the mathematical analogy would become a physical fact. Another motivation was experimental. It had recently been understood that the energy of an atom could take only certain special values. It had also been found that these 'energy levels' were related to each other by whole numbers (one energy value being, for example, exactly four times another). No one knew why. All this reminded de Broglie of the way musical notes—the frequencies of a vibrating string—took special values which were related to each other by whole numbers. To his mind this suggested that atoms must contain waves. In particular, if the orbiting electron in an atom was guided by a wave, a whole number of wavelengths would have to fit around the orbit. This could explain why only certain energies were allowed, and why the values were related by whole numbers.

De Broglie also understood very quickly that, in the microscopic domain, the laws of motion as laid down by Newton must be wrong (even with Einstein's modifications). His argument for abandoning Newton's laws was both simple and profound. It should rank as one of the great arguments of twentieth-century physics, on a par with the simple thought experiments involving light beams and moving observers for which Einstein is justly famous. Remarkably, to this day de Broglie's argument seems to have gone entirely unnoticed (except by this author).

De Broglie's argument is illustrated in Figure 41. A particle—in this case a photon or 'light quantum'—approaches a barrier with *one* hole. The particle is accompanied by a wave. When the wave hits the hole it undergoes 'diffraction': the hole acts like a source of circular waves, making the incoming wave spread out as shown in the figure.[20] What happens to the particle? De Broglie knew, from experiments with both ordinary light

---

[20] As the reader may recall from high-school physics experiments with water waves, this happens provided the wavelength is not much smaller than the width of the hole.

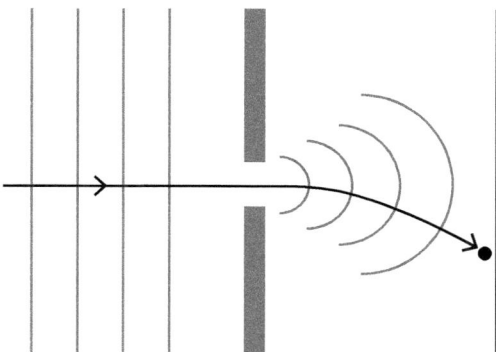

Figure 41. De Broglie's farewell to Newton's first law. A diffracting particle does not move in a straight line, even though there are no forces acting on it.

and X-rays, that the particle can be found at a point on the backstop lying above or below the hole. So far, so simple. But now, even without trying to watch the particle as it moves, we can make a simple deduction: the particle did not move in a straight line. The particle must have had some sort of curved trajectory, perhaps like the one shown in Figure 41. And now for the punchline. To our knowledge there are no forces acting on the particle. According to Newton's first law it should move in a straight line. And yet it does *not* move in a straight line. What is going on?

At this point most physicists would have taken a conservative route and concluded that, despite appearances, there must be some force acting on the particle, a new force we do not yet know about. De Broglie took a more radical route and proposed instead that Newton's first law must be wrong—and not only for photons. According to de Broglie, ordinary particles of solid matter are also accompanied by waves. And so they should also undergo diffraction. And so they must also violate Newton's first law. The most basic law of mechanics—what had arguably been the foundation stone of physics for several centuries (even with Einstein's modifications)—had to be thrown away.

In the seventeenth century Newton taught us that forces cause acceleration: if there is no force a body will move at constant speed in a straight line (think of a stone sliding on ice). Understanding this point had allowed physicists to abandon the ancient ideas of Aristotle—who thought that a force had to be applied to a body to keep it moving even at uniform speed.

But if we now follow de Broglie and abandon Newton's first law, the whole of Newton's physics collapses like a house of cards (as does Einstein's physics). This is not a step to be taken lightly, as de Broglie well knew.

In his PhD thesis of 1924 de Broglie proposed a radical rethink of mechanics. Instead of forces that cause particles to accelerate, as assumed by Newton and Einstein, in de Broglie's physics pilot waves cause particles to have a *velocity*. In a sense de Broglie was setting the clock back two thousand years and returning to Aristotle.

Such radical ideas are all very well, but do they make any testable scientific predictions? Yes, they do. In one of his first papers on pilot-wave theory, in 1923, de Broglie predicted that material particles like electrons would undergo diffraction and interference—something that no one else had remotely suspected. This seemed quite absurd to most physicists when they first heard it. How could ordinary solid matter behave like a wave?

De Broglie defended his PhD thesis at the Sorbonne in November 1924. The members of the examining committee were perplexed. There were interesting and radical ideas in the work, certainly. But was it all so much nonsense? The committee turned to an outside authority for an opinion. Not any authority, but the greatest and most original theorist of the age. Einstein's verdict was unequivocal:

He has lifted a corner of the great veil.

Einstein believed that de Broglie had taken a decisive and important step in illuminating the riddle of the quantum, a riddle which Einstein himself had struggled with since his light quantum (or photon) hypothesis of 1905. Einstein was inspired by de Broglie's thesis and encouraged others to read it.

In November 1925, Schrödinger wrote to Einstein that he had just

. . . read with the greatest interest the ingenious thesis of Louis de Broglie . . ..

Soon afterwards, in a series of papers published in 1926, Schrödinger found the correct equation for de Broglie's waves. This is now called the 'Schrödinger equation'—the basic textbook equation for the quantum wave. In a mathematical tour de force, Schrödinger applied this equation to calculate the energy levels of the hydrogen atom. His results agreed with experiment, in spectacular vindication of de Broglie's idea.

In the meantime, something extraordinary was happening in the lab. Beginning in the early 1920s two American physicists, Clinton Davisson and Lester Germer, were busy firing a stream of electrons at a crystal of nickel. The electrons would hit the surface and bounce off in various directions (Figure 42). The point of the experiment was to study the surface of the crystal. It may look smooth to us, but for a tiny electron the crystal surface is actually quite rough. So they expected the electrons to bounce off in more or less random directions (think of throwing a tennis ball at a jagged rock face). But the closer they looked, the more they noticed that the electrons seemed to be bouncing off in preferred directions—many came out at some angles, few came out at other angles. Why?

Something similar had already been noticed with X-rays. When they are fired at a crystal, X-rays bounce off in certain preferred directions. And everyone knew why: X-rays are waves. The regular structure of a crystal acts like a 'diffraction grating': any wave that bounces off it will come out in preferred directions. This was well known. But Davisson and Germer were firing *electrons* at the crystal, not X-rays, and everyone knew that electrons were particles. So why were the electrons coming out in preferred directions?

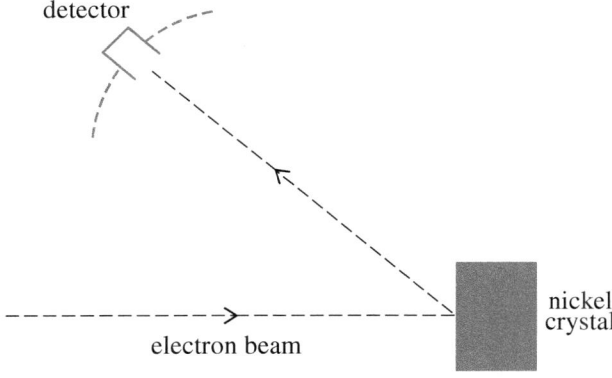

Figure 42. Electron diffraction as observed by Davisson and Germer in 1927. The electron is scattered in preferred directions, confirming de Broglie's theory.

As news of de Broglie's thesis spread, more people started to suspect that the electrons were being diffracted by the crystal as if they were waves. By 1927, the evidence was unequivocal. Electrons were not simply particles but 'matter waves'. And the data fitted de Broglie's theory perfectly.[21]

De Broglie was now famous. In 1927 a meeting was convened in Brussels, with the aim of bringing together the leading lights of the quantum revolution. Naturally, de Broglie was invited to present his ideas—along with Schrödinger, Heisenberg, and Born (who all had their own ideas, quite different from de Broglie's). This was the celebrated fifth Solvay conference, which we discussed briefly in Chapter 1. There de Broglie presented his final formulation of pilot-wave theory, this time not to a puzzled thesis committee in Paris but to a select group of the most distinguished theoretical physicists of the age, including Bohr, Einstein, Heisenberg, and Schrödinger.

De Broglie at the Solvay conference

What happened? According to standard historical accounts, de Broglie's theory was 'hardly discussed' at the conference. And when it was discussed, it was simply dismissed. The 'only serious reaction' came from the famously acerbic Austrian physicist Wolfgang Pauli, whose objection de Broglie was unable to answer. In any case, what de Broglie presented was quite primitive: it could only apply to one particle and not to a system of many particles. He was clearly out of his depth. After this withering defeat, de Broglie ran home with his tail between his legs and abandoned his misguided ideas. And that was that. What we now call quantum mechanics emerged victorious. Old-fashioned 'naïve realism' had been defeated. Or at least: that is what the history books more or less tell us.

It is all wrong, as is easily proved by visiting a good university library. There we find a copy of the Solvay conference proceedings which, as we have said, were published in French in 1928. Even with minimal French we quickly notice, just by perusing the table of contents, that de Broglie gave a lecture at the conference. So did Schrödinger. So did Heisenberg and Born, who gave a joint lecture. In this way three different versions of quantum

---

[21] Technically, de Broglie's theory predicts that the wavelength of the matter wave will vary inversely as the speed of the particle.

physics—three different theories—were presented.[22] As we thumb our way through the volume, we see that at the end of each lecture there is a discussion. And at the very end of the book there is a long 'general discussion'. In the discussions each participant is named—with a record of what they said, and of what others said in reply. It is all there, printed in the book. We do not have to rely on the historical accounts. We can see for ourselves.

So what did they say? Even with no French at all we notice that at the end of de Broglie's lecture there is a discussion that goes on for several pages. Why, then, does a widely cited book on the history of quantum physics, written by a distinguished historian, say that de Broglie's theory 'was hardly discussed at all'? In fact, if we amble our way through the final general discussion we see that de Broglie's name and ideas frequently pop up there as well. With a smattering of French and some knowledge of physics, it does not take much browsing to conclude that de Broglie's theory was discussed more or less as much as the other two theories. And there is more. As we flick through de Broglie's lecture, we come across his section 7 with the title 'La dynamique des systèmes', in which de Broglie applies his theory to systems of *many* particles. What is that doing there? The equations for what physicists call a 'many-body system' are there printed on the page for all to see—despite received opinion that de Broglie had a primitive theory for one particle only. De Broglie's reply to Pauli is also there in the general discussion, recorded for posterity. His reply was not perfect—but Pauli's objection was itself confused and misleading. Careful analysis shows that de Broglie's answer was essentially the right one.[23]

Why did the historians get it so wrong? Perhaps they misread the conference proceedings. As often happens, they may have seen what they already thought was there. Or perhaps they did not read them much at all. One thing is certain: instead of considering the published proceedings carefully, many historians relied on one-sided and misleading accounts of the conference written decades later by Bohr and Heisenberg—leading, as noted in Chapter 1, to the widespread skewed perception that the centrepiece of

---

[22] There were also two lectures on experimental physics by Bragg and Compton.
[23] Sceptics with no French can see the details for themselves—as noted in Chapter 1, a complete English translation, with detailed analysis and commentary, was published in 2009 by G. Bacciagaluppi and this author.

the conference was the so-called Bohr–Einstein debate, among many other misunderstandings.

De Broglie's lecture is a fascinating read. After summarising his thinking since 1923, de Broglie proposed what we now call pilot-wave theory. He argued that his 'new dynamics' had been experimentally vindicated by Davisson and Germer, and that it could explain all the strange phenomena of quantum physics. Among other applications, de Broglie described how pilot-wave theory could explain the wavy pattern seen in the two-slit experiment. He assumed that the incoming particles already have positions distributed according to Born's formula, and he then showed how the particles landing at the backstop will also be scattered according to Born's formula. In this way de Broglie was able to explain the characteristic wavy pattern, with each particle going through *one* hole only.

During the discussion after de Broglie's lecture, French theorist Léon Brillouin applied pilot-wave theory to the simple case of a photon bouncing off a mirror—and showed his audience a sketch of the trajectory (Figure 43).[24] Notice the characteristic wriggling motion close to the mirror, where the incident and reflected waves overlap. The waves form an interference pattern (like that seen in a two-slit experiment) and the particle avoids the 'dark fringes' where the wavy pattern is small.

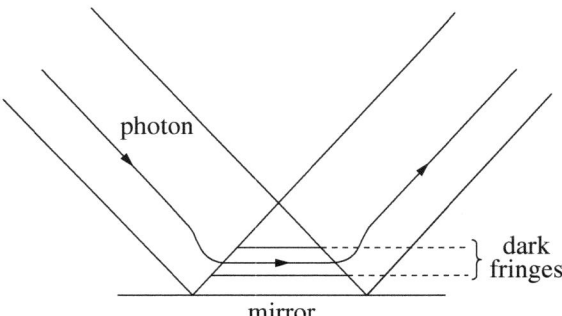

Figure 43. Quantum pornography in 1927. A particle trajectory according to pilot-wave theory, for a photon bouncing off a mirror, as drawn by Léon Brillouin at the 1927 Solvay conference.

---

[24] We now know that in pilot-wave theory photons, or particles of light, are best described in terms of fields. See the final section of Chapter 4.

And so we have come full circle: the 'quantum pornography' which Dewdney had shown to a shocked audience at a conference in Italy in 1985 had already been seen at the Solvay conference in Brussels in 1927.

Ironies of history

At this point the reader may well be puzzled. If de Broglie was able to explain the two-slit experiment, with a theory in which particles go through one hole or the other, then why did a scientist of Feynman's stature deliver lectures to undergraduates, decades later, telling them that such an explanation was impossible? And why are similar statements still frequently repeated in university physics courses all over the world?

The puzzlement deepens when we remember that de Broglie was the first scientist to predict that electrons would diffract and interfere. He made this prediction in 1923 from his new theory of particle motion. The prediction was vindicated four years later by Davisson and Germer. And yet, for the rest of the twentieth century and beyond, physicists have routinely cited electron diffraction and interference as experimental proof that electron trajectories cannot exist. How can something seen in the lab count as evidence *against* the theory that first predicted it? Something, somewhere, went terribly wrong.

Perhaps Feynman had never heard of de Broglie's theory, or of Bohm's revival of it? In fact, Feynman had extensive discussions with Bohm while on sabbatical leave in Brazil during 1951–2, when it appears he conceded that pilot-wave theory was a logical possibility (as recorded by Bohm in letters to a friend at the time). Bohm had also presented his work at a scientific conference, in Belo Horizonte, attended by Feynman. In a paper on pilot-wave theory, submitted for publication in June 1952 and published in January 1953, Bohm thanked Feynman 'for several interesting and stimulating discussions'. So Feynman clearly knew about Bohm's work first hand. And yet he went on to become one of the most prominent deniers that such theories were even possible.

As for the widespread failure to acknowledge de Broglie's ideas, this goes back to the 1920s. Even then hardly anyone knew what de Broglie had actually done. In 1929 de Broglie was awarded a Nobel Prize 'for his discovery of the wave nature of electrons', so his work was obviously regarded as important. But the historical evidence suggests that most physicists did not read de Broglie's papers or his thesis, and instead relied on second-hand reports. Only one aspect of his work entered the collective consciousness: that electrons could behave like waves. De Broglie's new theory of motion was not noticed, nor was his argument about the need to abandon Newton's first law. Even his basic idea that the wave guides the motion of the particle was ignored. In short, pretty much all of his work passed people by, except for the one point that the electron could somehow act like a wave. Much the same is true today: standard textbooks cite de Broglie only for the idea of matter waves (with a specific relation between speed and wavelength). His revolutionary rethink of mechanics and motion is not mentioned.

Why did most physicists not read de Broglie? There may have been a sociological factor. In the early twentieth century, France was something of a backwater in theoretical physics (though strong in mathematics). The main centres were in Germany, Austria, Denmark, and England. The other main architects of the quantum revolution—Bohr, Heisenberg, Born, Schrödinger, Pauli, Dirac—would often visit each other and corresponded regularly. De Broglie was something of a loner, working in isolation in Paris, pursuing his own thoughts and ideas.

But there is a more concrete reason why the spotlight did not linger on de Broglie: his idea of matter waves was taken up by Schrödinger, who ran with it. With scant public acknowledgement of de Broglie, Schrödinger reworked the analogy between optics and mechanics in a different way, so that today Schrödinger is often credited with insights that were really de Broglie's. Schrödinger's decisive contribution was to find the mathematical equation for de Broglie's waves, and to apply it to calculate atomic energy levels. De Broglie's idea about how to explain those energy levels turned out to be correct, but it was Schrödinger who worked out the details. This put de Broglie in the shade: why bother looking into de Broglie's work when we now had the Schrödinger equation?

The birth of the measurement problem

And so now we come to a decisive crossroads, a decision taken by one man that led physics into the deepest and darkest confusion. In his work of 1926, Schrödinger removed the particles from de Broglie's theory and kept only the waves. He believed, or at least claimed to believe, that only the waves were real, and that everything could be explained with the waves alone—even though there were obvious and devastating objections. Waves tend to spread out over space. They do not stay confined to a small region. But when we look at a particle we find a tiny object confined to a small region. How can we explain what we see if there is only an extended wave?

That this was problematic did not go unnoticed. Here, for example, is Pauli, writing to Bohr in August 1927, commenting on de Broglie's recent work on pilot-wave theory and its contrast with recent work by Schrödinger:

> In the last number of the Journal de Physique, a paper by de Broglie has appeared . . . it is very rich in ideas and very sharp, and on a much higher level than the childish papers by Schrödinger, who even today still thinks he may . . . abolish material points.

If there were only waves and no material points, it was difficult to explain why we see a tiny particle-like object when we make a measurement.

This objection is the seed of a larger problem. When we make a measurement in quantum mechanics we find only one result—even though the quantum wave is usually a 'superposition' (or combination) of many possibilities. For example, think of a wave hitting the backstop after passing through a barrier with only one hole (Figure 44). The wave is spread all over the backstop. And yet, each time we perform the experiment, we find the particle has landed at only *one* point. How can we explain that? In pilot-wave theory it is quite simple: the particle starts at some initial point, follows a trajectory guided by the wave, and lands at some final point (Figure 41). But if instead there is no moving particle and just the wave, how do we end up with a small blip at only one point? This very question was posed by Einstein, in his one significant contribution to the (public) discussions at the 1927 Solvay conference. Einstein suggested that de Broglie was right to include the moving particle as well as the wave, but his argument was not widely understood.

Thus was born the notorious 'measurement problem', which has bedeviled theoretical physics for nearly a century. As we saw in the Prologue, the problem is often presented dramatically with the example of Schrödinger's cat, which seems to be both dead and alive. But we have the same problem for a single particle: the equations seem to say that the particle exists simultaneously at many different points in space, even though when we

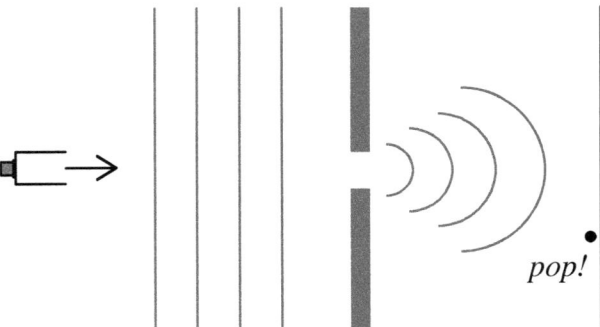

Figure 44. Popping into existence. The madness of the measurement problem.

make a measurement we find it at only one point. Does the particle simply pop into existence—when we choose to look at it?

To see how unclear the quantum textbooks are, let us take a closer look at the diffraction experiment now shown in Figure 44. This is how university physics students are taught to think about it. A single particle is fired at the barrier. But there is no moving particle, there is *only* a wave. So, as shown in the figure, a wave comes in and hits the barrier, diffracts at the hole, and spreads all over the backstop. Then what happens? Well, when we look at the backstop—*pop!*—a particle appears, seemingly out of nowhere. Why did it appear only when we looked at it? Because, in quantum physics, reality depends on the observer. And why did the particle appear there and not elsewhere? For no reason. We are witnessing the fundamental randomness of the universe. Convinced?

The quantum measurement problem has prompted many attempts to solve it. Some say that the Schrödinger equation must be modified so that the wave suddenly and randomly collapses to a point—giving the appearance of a tiny blip somewhere on the backstop. In other words, the wave suddenly contracts to a small region, making it look as if a particle has appeared. This idea was pioneered by the American physicist Philip Pearle in the 1970s, and has since been advocated by (among others) British Nobel Prize-winning theorist Sir Roger Penrose. Others say that we should keep the Schrödinger equation as it is—so the wave always spreads out—and take seriously the idea that the particle exists in many different places at the same time. This is the 'many-worlds' or parallel-universes interpretation of quantum mechanics, which was pioneered by another American physicist, Hugh Everett, in his PhD thesis of 1957, and which has since been advocated by (among others) the distinguished British physicist David Deutsch. According to this interpretation, when the wave spreads out the whole universe divides into many different branches: in one universe we see the particle here, in another universe we see the particle over there, and so on. Both ideas—collapse theories, many worlds—are serious attempts to solve the measurement problem, and both have a significant following among physicists today.

But wait a minute. If Schrödinger had not removed the particles from de Broglie's theory, there would not *be* a measurement problem. A particle would simply land somewhere depending on where it started from (as in Figure 41). In that case, there would be no need to change the Schrödinger equation to make the wave collapse to a point, and no need to believe in parallel universes. Schrödinger arguably made a mistake, which *created* the measurement problem—and the burning need to solve it.

This brings us to the ultimate irony. Some thirty years later, after Bohm had revived and extended de Broglie's theory, Bohm was accused by Heisenberg of adding an 'ideological superstructure' to quantum mechanics. Since then, the few physicists who work on pilot-wave theory have faced similar accusations: that they are adding excess structure, with no rhyme or reason or justification. For example, commenting on pilot-wave theory in 1986, Deutsch claimed that it is almost certainly a mistake

. . . to append to the quantum formalism an additional structure . . . solely for the purpose of interpretation . . . without any physical motivation.

But nothing is being 'appended'. In pilot-wave theory we are just including what was already there and (arguably) should never have been removed. As for physical motivation, de Broglie had plenty: from the behaviour of X-rays, to the analogy between optics and mechanics, and the puzzle of atomic energy levels.

To this day many commentators portray pilot-wave theory as if it were somehow cooked up to solve the measurement problem. But pilot-wave theory was there before the measurement problem. In fact, it is fair to say that the measurement problem was created, artificially, by Schrödinger: by removing the particles from de Broglie's theory, he condemned theoretical physics to a century of needless controversy and confusion.

The trouble with pilot-wave theory

Pilot-wave theory is surely one of the most misunderstood theories in the history of physics. As we have seen, from the beginning hardly anyone knew what de Broglie was really doing and few took the trouble to read his PhD thesis of 1924. The one time his ideas got a fair hearing, at the top table of theoretical physics, was at the 1927 Solvay conference. But even on that occasion, what actually happened was swept under the rug by historians. Since then and right up to our own time, pilot-wave theory has been subject to repeated misunderstandings, distortions, and even a certain kind of mysticism.

To start with, by 1930 de Broglie himself had turned away from his own creation. Not because he did not understand it, as many have claimed, but mainly because he understood it all too well. De Broglie saw that pilot-wave theory pointed to a reality grounded in a higher-dimensional configuration space. This was a conclusion he could not fathom or accept. Like most physicists he believed that reality had to be grounded in our everyday three-dimensional space. Configuration space was a mathematical fiction, hence so was the pilot wave, and so was pilot-wave theory. In retrospect this was a mistake. We have seen that a higher-dimensional reality can help us understand nonlocality. For that reason alone it is worth taking seriously. But in 1930 nonlocality itself was not taken seriously. It was not until Bell's work of 1964 that humanity slowly began to wake up to the idea that it lived in a nonlocal universe. In a sense de Broglie was too far ahead of his time. By a remarkable mix of imaginative leaps and scientific reasoning, he had arrived at a place that seemed impossibly unreal. He shrank back.

We know that Einstein was one of the few who actually read (and deeply appreciated) de Broglie's PhD thesis. Not surprisingly, at the 1927 Solvay conference, Einstein voiced support for de Broglie's ideas. But in the meantime Einstein had been up to something: a few months before the conference he had tried to develop his own version of pilot-wave theory. Remember that in de Broglie's theory the pilot wave guides the motion of particles according to a simple law: the particles move perpendicular to the wave crests, with a speed that goes inversely as the wavelength. Einstein proposed an alternative law that was much more complicated.[25] Why he did this is unclear.

Einstein was unparalleled in his ability to discern theoretical simplicity and elegance. And yet, on this occasion, his alternative pilot-wave theory was strangely overcomplicated and clunky. Worse, as shown much later by Holland in 2005, in many circumstances Einstein's equations did not even make mathematical sense.[26] We do not know if Einstein noticed these problems. What we do know is that he was about to publish his theory in a journal but withdrew it at the last minute in a hasty phone call to the editor. The original manuscript survives in archives held at the Hebrew University of Jerusalem. We also know that Einstein had been planning to present something at the Solvay conference. His aborted version of pilot-wave theory was probably going to be it. Regretfully he withdrew his commitment to speak, telling the organisers that he had 'tried with all my strength' to prepare something 'of value' for the conference but had now 'given up'. Einstein knew about de Broglie's simple law of motion, so why did he choose an alternative that in retrospect seems so bizarrely baroque by comparison—as well as problematic? No one really knows.

A quarter of a century had to pass before de Broglie's ideas were resurrected—by Bohm in 1952. We have said that in some ways Bohm extended the theory. He clarified how it explains the general theory of quantum measurement. And he applied it to the electromagnetic field. These were important advances. But in other respects Bohm misunderstood the theory. He thought of it in terms of Newton's laws and Newtonian forces (ideas which de Broglie had rightly abandoned at the outset). Bohm imagined there was an extra 'quantum force' accelerating the particles—in order to explain, for example, the curved trajectory in Figure 41. It was possible to put it like this mathematically, but the resulting equations looked artificial. Bohm also thought of the pilot wave as being a bit like an electromagnetic force field. He did not see, as de Broglie had done, that the pilot wave was a radically new kind of thing. All in all, Bohm made the

---

[25] Technically, Einstein proposed that the pilot wave determines a 'curvature' in configuration space, which in turn determines the trajectories according to what now seems a rather odd mathematical construction.

[26] The velocity field defined by Einstein can take imaginary values. In addition, Born's formula is not preserved over time.

theory seem much more like classical physics than it really was, disguising the conceptual advances that de Broglie had made. Bohm was looking at a radically new kind of physics through the eyes of Newton. De Broglie's original vision, of a new and non-Newtonian theory of motion, had been lost.

We now know that Bohm's Newtonian version of pilot-wave theory is scientifically untenable. The reason has to do with quantum relaxation. In de Broglie's theory, we can explain quantum noise as having come about by the complicated wriggling motions of the particles, which end up being spread around in space according to Born's formula (Figure 23). We might hope for a similar explanation in Bohm's version, but it does not work. In Bohm's theory both the initial positions *and* the initial velocities are unconstrained by any law (just like in Newton's physics). If we start with an anomalous distribution of positions and velocities, we find that there is no relaxation to Born's formula. In fact, small deviations from Born's formula can quickly grow and become very large. Technically, Bohm's version of the theory is 'unstable'—a devastating conclusion that was understood only in 2014.[27] The long-standing disagreement over de Broglie's original theory versus Bohm's later version is no longer a matter of taste or theoretical preference: Bohm's version is actually wrong.

The vexing question of Born's formula

While Bohm's version of the theory cannot explain quantum relaxation and Born's formula, in some passages of his papers of 1952 Bohm had sketched ideas not dissimilar to those proposed in this book: that quantum noise should be understood as analogous to a state of thermal equilibrium. A year later, in response to criticisms (including from Pauli) that Born's formula needed an explanation, Bohm studied an example of a molecule colliding repeatedly with other particles and argued that the molecule would relax to the equilibrium state described by Born's formula. Bohm's argument could work because he kept the velocities fixed by de Broglie's law—so that he was, in effect, using de Broglie's version of the theory (though seemingly without realising it). Even so, Bohm did not provide a general argument for relaxation. In 1954, Bohm returned to the theme together with French physicist Jean-Pierre Vigier (who was then de Broglie's assisant in Paris).

---

[27] In work by this author in collaboration with the Belgian physicist Samuel Colin.

Citing difficulties with understanding relaxation, Bohm and Vigier proposed that the theory should be modified by adding a randomly fluctuating subquantum 'fluid', whose purpose was to make the particles jump around at random in such a way as to explain relaxation to Born's formula. This amounted to abandoning the strict 'determinism' of pilot-wave theory, in which every effect has a cause.

In retrospect, this last move by Bohm and Vigier seems an unfortunate mistake. Only a year or two earlier, Bohm had been within striking distance of understanding what we now know to be true: that quantum relaxation can be explained by the original 'deterministic' theory (in de Broglie's version), with strict laws of cause and effect, and with no need for any underlying randomness. On the other hand, had Bohm persisted with his Newtonian version, it would never have been possible to understand relaxation, and a move to a theory with underlying randomness would have been necessary. Still, with hindsight, it seems a pity indeed that Bohm came so close to what we now know about quantum relaxation, and then drew back.

Among the small number of theorists who promoted pilot-wave theory after the 1950s, Born's formula was taken as a law or axiom. The original idea of quantum relaxation was lost and forgotten—until its revival by this author, in two papers published in 1991. These papers gave a general understanding of quantum relaxation, based on de Broglie's original version of pilot-wave theory, an understanding which has since been backed up by extensive computer simulations.[28] It was also shown how violations of Born's formula would allow superluminal signalling and a breakdown of the uncertainty principle, potentially opening up a new domain of physics beyond quantum mechanics. It was argued that Born's formula describes a state of quantum death, analogous to heat death, which probably arose by relaxation in the early universe. This line of thought was the basis of the author's PhD thesis and is the main subject of this book.

A year later, in a paper published in 1992, an alternative approach to understanding Born's formula in pilot-wave theory was proposed by German theorist Detlef Dürr, American theorist Sheldon Goldstein, and Italian theorist Nino Zanghì. This group also adopted de Broglie's original

---

[28] Technically, quantum relaxation can be understood in terms of a coarse-graining H-theorem, analogous to that commonly used to understand thermal relaxation. Computer simulations show that, in appropriate circumstances, the coarse-grained H-function (which measures the distance from equilibrium) decreases approximately exponentially with time.

version of the theory but, rather confusingly, renamed it 'Bohmian mechanics' (a misnomer which disregards de Broglie's priority and misrepresents Bohm's work). On their understanding, Born's formula is hard-wired into pilot-wave theory—we must be, always have been, and always will be, confined to a state of quantum death. Quantum noise and the uncertainty principle are unavoidable. As they put it, in a universe governed by pilot-wave theory, there are 'irreducible limitations on the possibility of obtaining knowledge' and 'absolute uncertainty arises as a necessity'. What is the basis for these remarkable claims? In their paper, Dürr et al. *assume* that Born's formula is true for the whole universe at the beginning of time, and from this they deduce Born's formula for individual particles at later times. We are then supposed to conclude that violations of Born's formula are essentially impossible. But the argument assumes what it sets out to prove—that Born's formula is a fundamental law of probability. This circular reasoning has been influential among some philosophers commenting on the theory, but physicists active in the field are usually unconvinced.[29] As we will see in this book, whether Born's formula is violated in our world (past or present) is ultimately an empirical question, to be settled by experimental evidence and not by theoretical or philosophical decree.

At issue here is a basic conceptual distinction between initial conditions on the one hand and laws of physics on the other. Recall our example of throwing a tennis ball (Chapter 2, Figure 33). As we discussed, where the ball lands depends on its initial position, speed, and direction of travel (at the time it leaves our hand). Those are the initial conditions. While we did not say so explicitly, of course where the ball lands also depends on the operative laws of physics—in this case Newton's laws of motion, which determine the trajectory of the ball, given the initial conditions. Now the laws are immutable: they are always the same and cannot be any different from what they are. In contrast, the initial conditions could have been different. We could, for example, have thrown the ball from a different position or with a different speed. There is no law that fixes the initial conditions. They are, we might say, contingencies. So there is a deep distinction between initial conditions, which can vary, and laws of physics, which are always the same.

---

[29] For a detailed rebuttal—and an extensive critique of the 'Bohmian mechanics' school generally—technical readers are referred to a paper by this author published in 2020.

With that point in mind, we can understand what went wrong with the general understanding of Born's formula in pilot-wave theory. Once the idea of quantum relaxation had fallen by the wayside, people in the field started to think of Born's formula as a law of physics, which cannot vary—when really it is an initial condition, which can certainly vary. Taking Born's formula as a law is like insisting that the positions of tennis balls must always be spread out or scattered in a particular way (that is, must have a particular distribution), when in fact the positions of the balls are arbitrary and unconstrained. The basic distinction between initial conditions and laws of physics was clear enough in Bohm's papers of 1952, but was later lost. The confusion was compounded in 1992, when Dürr and co-workers tried to argue that, somehow, the laws of physics require us to apply Born's formula to the beginning of the universe. But just as tennis balls can start from any location, in pilot-wave theory the initial conditions of the universe *need not* obey Born's formula, a subject we will discuss in detail in Chapter 6.

To consolidate the key point as clearly and simply as we can, let us return to our analogy between throwing a tennis ball (Figure 33) and firing a particle at a two-slit barrier (Figure 30). Once the tennis ball is in the air, its motion is determined by the immutable laws of physics—but precisely where the ball started from is not determined by any law, and indeed the ball could have started from elsewhere. Similarly, in pilot-wave theory, once the particle has left the source, its motion is determined by the immutable laws of physics—but precisely where the particle started from is not determined by any law, and (again) indeed the particle could have started from elsewhere. If these experiments are repeated many times, the probability distribution or scatter of the starting positions (whether of tennis balls or of particles) is not determined by any law but is, as a matter of principle, quite arbitrary—for the initial particles we might have Born's formula (as in Figure 32), or we might not (as in Figure 34).

These fine distinctions may seem arcane, but the widespread failure to appreciate them has seriously damaged the field. Today pilot-wave theory tends to be ignored by most physicists, and for understandable reasons. As it is usually presented, with Born's formula as a law or axiom, the theory contains no observable new physics. Its predictions are always the same as those of quantum mechanics. So why bother with it? For as long as quantum noise and the uncertainty principle hold sway, the trajectories described by de Broglie and Bohm can never be observed and their details can never be tested or confirmed. With some justification, then, most scientists are inclined to doubt those trajectories really exist, and to them the theory seems rather pointless. But this is all a tragic misunderstanding. In pilot-wave theory, Born's formula is not a law but merely a particular initial condition. Born's formula can be broken—in which case the underlying physics will be revealed, including the details of the trajectories. Much of the rest of this book tries to explain how we can, at least in principle, escape from quantum death and evade the uncertainty principle—and thereby gain access to the radically new physics of quantum life that is currently beyond us.

At this point, it will be instructive to backtrack a little in time, and pick up the thread of Bohm's work and ideas. We left Bohm in 1954, with Vigier, abandoning determinism (cause and effect) in order to explain relaxation to Born's formula. We said that this was an unfortunate mistake. What did Bohm do next?

Quantum mysticism: wholeness and nonlocality

Beginning in the 1960s, Bohm's thinking took a decidedly mystical turn. He developed an interest in ancient Indian thought, as expressed for example in the *Yoga Sūtras* of Patañjali (written nearly two thousand years ago). Also discernible is the influence of German philosophical idealism, and of Alfred North Whitehead's 'process philosophy' (where reality is said to consist not of things but of processes). Bohm's new outlook culminated in a book, *Wholeness and the Implicate Order*, published in 1980. Among other ideas Bohm suggested that we develop a new form or mode of language, which he called the 'rheomode' (from the Greek 'to flow'), where the traditional grammar of subject, verb, and object is supposedly superseded. Bohm claimed that our current use of language makes us think in terms of static and fragmented things, when instead reality is an undivided and ever flowing 'process'. For the later Bohm, the usual distinction between mind and matter had to be jettisoned. Conscious thought, or intelligence, was part of the primal stuff of reality—a reality that was ultimately unfathomable and mystical. As Bohm put it:

> Intelligence and material process have thus a single origin, which is ultimately the unknown totality of the universal flux.

During this period Bohm also had a long-standing association with the Indian philosopher-guru Jiddu Krishnamurti, with whom he published a book based on their dialogues. Bohm had come a long way since the 1950s, when his outlook had been largely rooted in Newtonian physics, with hints of Marxist-Leninist ideology in the background.

We can now understand why, by the 1970s, Bohm was no longer interested in pilot-wave theory. In 1977 our pioneering PhD student Chris Dewdney had arrived at Birkbeck College in London to work in Bohm's research group on the foundations of quantum physics. But Bohm did not even mention his work from the 1950s. Almost a whole year passed by before Dewdney learned about pilot-wave theory—and not from Bohm, but from a book he came across by chance while browsing in a nearby bookshop. Published in 1973 by the Dutch physicist Frederik Belinfante, the book was a survey of 'hidden-variables theories' and had a chapter devoted to Bohm's work of 1952. Dewdney was astonished and wondered why Bohm had not said a word to him about this work. Bohm was in turn startled to see the pictures produced by Dewdney's calculations. It was as if Bohm no longer appreciated what he had achieved twenty-five years earlier. Dewdney sensed that Bohm had become discouraged by the poor response to his earlier work, which tended to be either misunderstood or simply ignored. In any case, Bohm was now absorbed by a certain kind of philosophical mysticism, which he was trying to relate to physics with little success.

When Dewdney rekindled his interest in pilot-wave theory, Bohm started to look at the theory through the lens of his new outlook. The later Bohm's unusual interests probably helped him to see more clearly what a radically new kind of physics pilot-wave theory really was. Bohm's interest in 'wholeness' encouraged him to emphasise nonlocality and to embrace it as a fact of nature, at a time (the 1980s) when most physicists still thought it beyond the pale. His keen interest in 'non-mechanistic' ways of thinking no doubt also inspired him to discern a profound feature of pilot-wave theory which others had passed over without comment. In conventional physics the universe can be divided into pieces that interact according to fixed laws. For example, in Newton's theory of gravity, a piece of matter here pulls on a piece of matter over there, with a force varying as the inverse square of the distance. Regardless of where we are or the circumstances, the force always goes as the inverse square of the distance. But in pilot-wave theory there are no fixed interactions between different parts of the universe: the way things interact depends entirely on what the pilot wave happens to be. This is radically different from any previously known physics, as Bohm now rightly emphasised. Despite these insights, however, Bohm continued to believe that pilot-wave theory could be at best only a crude step towards a deeper theory, one in which mind and matter were somehow combined.

Perhaps unsurprisingly, in the background there lurked an interest in the paranormal—in particular psychokinesis (mind over matter). Bohm shared this interest with his colleague at Birkbeck College, the experimental physicist John Hasted. One late afternoon in February 1974, Bohm and Hasted visited the illusionist, magician, and self-declared psychic Uri Geller, at his hotel suite in London, to observe a demonstration of his powers. As reported by Hasted, with Bohm and others as witnesses, Geller appeared able to bend metal keys by stroking them with his finger. A few months later, in laboratory tests at Birkbeck, Geller reportedly bent a small metal disc without touching it, and was able to trigger a Geiger counter—purely, it appeared, by the power of his mind. In the face of scathing criticism, including demonstrations by professional magician James Randi that these were mere conjuring tricks, in 1975 Bohm and Hasted hit back in the pages of the scientific journal *Nature*, claiming that the presence of sceptical observers in the lab would be counterproductive because 'the entire process goes most easily when all those present actively want things to work well'. Readers familiar with the history of psychical research—whose origins lie in the nineteenth-century fascination with spiritualist mediums supposedly communicating with the souls of the dead—will recognise the long-standing hallmark of a notoriously fraudulent field: the phenomena 'occur' only when all those present are eager to believe.

Bohm's interest in the paranormal was not as unusual as it may sound, however, when viewed in the broader historical context of British academia. As noted by the British historian Joanna Bourke, an interest in spiritualism and the paranormal had long been something of a tradition at Birkbeck. In fact, such interests were not uncommon among British academics in the nineteenth and twentieth centuries, especially in Cambridge, and in particular at Trinity College—several of whose fellows, including Nobel Prize-winning physicists Sir J. J. Thomson and Lord Rayleigh, were prominent members of the Society for Psychical Research (founded in 1882). This tradition continues down to our own time, with cosmologist Bernard Carr and Nobel Prize-winning theorist Brian Josephson.

At this point it is sobering to recall the solid experimental evidence that motivated de Broglie in 1923—atomic energy levels and the diffraction of X-rays—as well as his striking idea that these could be explained by combining the physics of particles with the physics of waves. This was theoretical physics at its best: an ability to discern which experimental puzzles were the most significant, and the imagination and insight to explain them. And yet, some fifty years later, the resulting theory had come to be associated with Eastern mysticism, the amalgamation of mind and matter, and the paranormal.

This was not the first time that quasi-religious mystical beliefs were read into the latest physical theories. Bizarre as it may seem today, in the nineteenth and early twentieth centuries it was suggested by some prominent physicists that the invisible and intangible electromagnetic field (or 'luminiferous ether' as it was then called) was the gateway to, or even the seat of, the human soul. Scottish physicists Balfour Stewart and Peter Guthrie Tait had respectively done important work on terrestrial magnetism and thermodynamics. In their book *The Unseen Universe* of 1875, to combat what they saw as a rising tide of atheism, Stewart and Tait argued for an invisible universe, which was the true realm of the spirit and of the Christian religion, and which was connected to the visible material universe by means of the ether. Prominent English physicist Sir Oliver Lodge had done important work on the detection of radio waves. In his book *Ether and Reality* of 1925, Lodge argued that human consciousness resided in the ether itself, with no need for a material body to sustain it. As he put it, the ether is 'the primary instrument of Mind, the vehicle of Soul, the habitation of Spirit'. According to Lodge, the ether is the repository of the 'spirit world' inhabited by the deceased—who, as he claimed in an earlier book, communicate with the living during spiritualist séances. Despite these ideas, today when we tap into the electromagnetic field by switching on the radio, we hear only the voices of living radio presenters and not the voices of the departed.

What is it about the unknown that can trigger the mind to mysticism? There seems to be a tension in the human spirit, pulling towards clarity on the one hand, and towards obscurantism on the other. Sometimes these traits are combined in the same person, or manifest at different stages of their lives. The early Bohm sought to bring clarity to quantum mechanics. In contrast, in some respects, his later work seems less an attempt to overcome obscurantism than a capitulation to it.

And yet, perhaps we should not too hastily dismiss the role certain kinds of ideas can play in the development of scientific thought. Bohm's later mystical interests arguably inspired him, as we have said, to consider non-locality as a real possibility, as well as to appreciate the non-mechanistic nature of pilot-wave theory, despite considerable ridicule at the time. In similar fashion, some 300 years earlier, Newton's well-known interests in alchemy arguably inspired him to consider gravitational attraction between the earth and the Sun—across empty space, with no intervening medium—as a real possibility, again despite considerable ridicule at the time.

In the seventeenth century, the prevailing 'mechanical philosophy' held that all scientific explanations—for physical and chemical effects, including the gravitational pull of the Sun on the earth—had to be based on local action by contact. In plain English this meant that one piece of matter could affect another only by touching it and giving it a push, in other words by direct contact, a view widely associated with the French mathematician, scientist, and philosopher René Descartes. According to these ideas, the Sun could act on the distant earth only indirectly, by means of invisible matter filling space. In Descartes' 'vortex' theory, the invisible matter supposedly swirled around the Sun and swept the earth and other planets along with it. Newton's claim that the Sun could act on the earth and other planets directly, across empty space, sounded like magic to many of his contemporaries, and was widely derided as a throwback to earlier occult thinking. According to one of Newton's most respected biographers, the American historian of science Richard Westfall, Newton's detailed and extensive study of alchemy eventually led him to reject the prevailing mechanical philosophy, and to suggest that matter had intrinsic (or occult) properties or powers whereby it could exert a force directly on other matter even across empty space.

Despite Newton's conceptual heresy, the mathematical and predictive power of his theory of gravity ensured that it reigned supreme for nearly 250 years. In 1915 his theory was radically modified by Einstein, in a way that removed action at a distance. Even so, Newton's theory was arguably the right approach at the time, and more generally his idea that physical and chemical effects should be explained by particles acting directly on each other without actually touching proved to be immensely fertile.

In the case of pilot-wave theory, we have an action at a distance far more severe than anything contemplated by Newton. Instead of decreasing with distance (as the inverse square), the nonlocal action between entangled particles remains the same *no matter how far apart* they may be. This is spooky by anyone's standards, and many physicists today respond to this idea with as much scepticism as the mechanical philosophers of Newton's time did to his idea of gravitational attraction.

To see the world as it really is often requires imagination—which can have decidedly unscientific sources. According to the American sociologist and historian Lewis Feuer, as well as the scientific motivations we have described, de Broglie too took inspiration from what might now seem a questionable source—the French philosopher Henri Bergson, whose writings against traditional mechanistic science were all the rage in the Paris of de Broglie's youth.

Be that as it may, at the end of the day, whatever the original motivations may have been, and however palatable or unpalatable we may find some of those motivations, scientific theories must ultimately be judged by cogent arguments and concrete evidence. It is time, then, to evaluate the evidence for nonlocality.

Action at a distance: fact or fantasy?

Today the scientific evidence for nonlocality is strong—though not without controversy. It is worth reviewing the evidence in some detail, as it provides the motivation for much of this book.

Our story begins in 1935 when Einstein, together with physicists Boris Podolsky and Nathan Rosen, published a paper arguing that there must be a deeper reality behind quantum mechanics—a hidden world of cause and effect, in which everything happens for a reason. Specifically, there are real quantities of which we are oblivious, and which determine the outcomes of seemingly random quantum events. Today we call such quantities 'hidden variables'. A theory containing hidden variables is called, appropriately enough, a 'hidden-variables theory'. Pilot-wave theory is an example: the precise (unknown) initial particle positions determine the outcomes of quantum events. Those initial positions are the hidden variables. But we are getting ahead of ourselves.

How did Einstein, Podolsky, and Rosen—or 'EPR' as they came to be called—deduce that there must be a deeper reality behind quantum mechanics? In itself the argument is quite simple. In fact, essentially the same argument had already been made earlier by Einstein, at the 1927 Solvay conference, but his reasoning was not understood. When a more elaborate version of the same argument was published in 1935, Bohr wrote a notoriously obscure reply.

To present the EPR argument, we need a bit of background. Consider first a single electron. If the electron passes through a magnetic field, it will be deflected upwards in the direction of the field or downwards in the opposite direction (Figure 45).[30] In quantum mechanics this is usually called a measurement of the electron 'spin' (or equivalently, of the electron 'magnetic moment'). Depending on the outcome we say that the electron has been measured to have 'spin up' or 'spin down' along a certain axis or direction. For our purposes, what matters is that the electron moves upwards or downwards along the axis specified by the magnetic field. Quantum mechanics tells us that individual outcomes—up or down—are random, with probabilities given by Born's formula.

Now consider a *pair* of entangled electrons, which are widely separated. Let each electron pass through a magnetic field. If the two magnetic fields point in the same direction, we can say we are measuring the spins of both electrons along the same axis. What happens? Well, that depends on the initial (joint) quantum wave. Let us assume that the initial quantum wave is the 'singlet state' (to use the jargon). In this particular state, and assuming the magnetic fields to be aligned, experimentally we find that the two spins

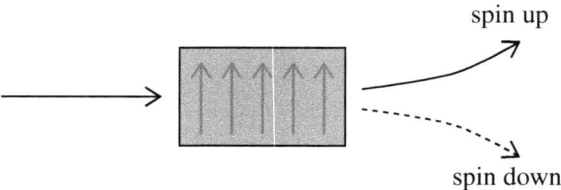

Figure 45. A quantum spin measurement. When an electron enters the magnetic field, it is deflected upwards or downwards, giving an outcome 'spin up' or 'spin down'.

---

[30] Technically, the magnetic field has to be non-uniform, and it is really the field gradient that determines the relevant direction. Also, in practice this discussion applies to an electron bound inside a neutral atom. But we can ignore these details here.

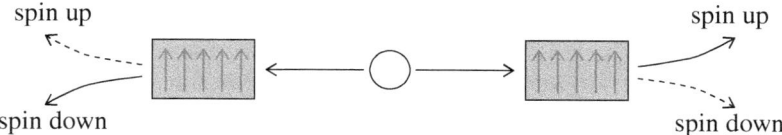

Figure 46. Spin measurements for an entangled electron pair, with magnetic fields pointing along the same axis. In the singlet state, the outcomes are always opposite—up and down, or down and up—no matter how far apart the electrons may be.

always come out opposite. In other words, if one electron moves upwards the other moves downwards, and vice versa (Figure 46).

We can now present the EPR argument. The point is so simple as to be almost trivial—blink and you might miss it. It goes like this. Imagine that, in Figure 46, one of the electrons is here in the lab, while the other is far away in a separate lab run by a friend. The initial quantum wave is the singlet state, for which the outcomes always come out opposite, as many experiments have shown. So, now, we pass our electron through a magnetic field and let us say it happens to come out 'up'. Our faraway friend has not yet passed their electron through their magnetic field (perhaps the electron has not reached the other lab yet). Even so, *we already know* that their outcome will be 'down' (since the outcomes always come out opposite for this particular quantum wave). Similarly, if our outcome happens to be 'down', we already know that their outcome will be 'up'. If we pause awhile and ponder, we may find our minds wobbling a little. Quantum mechanics, as usually understood, says that the outcome perceived by our friend is determined purely by chance, and comes into being only when their measurement is actually performed. If our faraway friend has not even done their measurement yet, how can *we* possibly already know what their outcome is going to be? It sounds suspiciously as if, really, the distant outcome is not a matter of chance at all but is somehow already determined or fixed. What is going on?

It might be thought that measuring the spin of our electron must somehow affect the other electron, forcing it to give an outcome that is always opposite to ours. To discount this possibility, EPR appealed to locality. In 1935 it was widely believed that the speed of all physical interactions is bounded by the speed of light, as required by Einstein's theory of relativity, and as is still widely believed today. This means that, for our widely separated electrons, what we do over here cannot instantaneously affect anything over there. In particular, if we send our electron through a magnetic field, this can have no influence on the other electron (if it is sufficiently far away). In other words, our measurement here can have no effect on the outcome of a measurement over there. And now for the crucial point. If we measure our spin here, we can predict the spin over there. But what we do over here can have no effect on anything over there. Therefore, the spin over there must be *already determined*—that is, the outcome is already fixed in the future—regardless of what we do (or do not do) over here. In particular, even if we do nothing over here, and do not bother to measure our spin, the spin over there must already be determined.

We can make a similar argument the other way around. If our faraway friend measures their spin, they can predict in advance what our outcome will be, even before we have made our measurement, again leading to the conclusion that the outcome of our measurement is *not* a matter of chance but is already determined—even if our faraway friend makes no measurements at all.

At this point we should pause and take in what appears to be a stunning conclusion. We seem to have deduced that, at least for this particular quantum wave, the outcomes of spin measurements for both particles are not random events, with probabilities given by Born's formula. Instead, each and every outcome is already determined (fixed in the future) before any measurements are made. But quantum mechanics does not provide any means or mechanism for individual outcomes to be determined; it gives only probabilities. Therefore, something must be missing from quantum mechanics, something we do not know about but which, in fact, determines the outcomes of measurements for each individual electron. In other words: there is a deeper level of physics containing hidden variables. As EPR put it, quantum mechanics is 'incomplete'.

That was in 1935. As we said, Bohr wrote a notoriously obscure reply. His counter-argument amounted to denying the existence of an objective reality before measurements are performed. It is not clear whether many physicists felt they really understood Bohr's reply or subscribed to it. Perhaps some took comfort in the mere fact that there had been a reply: someone senior has done the hard thinking for us, and we can get on with publishing papers about the practical applications of quantum mechanics. Not everyone took this view. Some physicists recognised the importance of the EPR argument, amid recurring controversy over its implications. Einstein had an uncanny ability to make simple physical arguments that penetrated deeply, even if they were not always appreciated or understood by his peers.

The EPR argument was, however, only the first of three milestones along the remarkable road that led to our present understanding of nonlocality.

The second milestone was Bohm's revival of pilot-wave theory in 1952, together with his clear understanding that the theory was nonlocal—a measurement done here could instantaneously affect the outcome of a faraway measurement over there. This was a 'completion' of quantum mechanics, with hidden variables that determined the outcomes of measurements, and so in that sense it was a theory along the lines envisaged by EPR. But it was nonlocal, contrary to what EPR had assumed. So where did this leave us?

We now come to the third and final milestone, with the publication in 1964 of Bell's groundbreaking paper on what we now call 'Bell's theorem'. By his own account, Bell had read Bohm's papers when they first appeared, some twelve years earlier, and had been duly impressed by the clarity and explanatory power of pilot-wave theory. But he was disturbed by its non-locality, which runs counter to what every modern physicist is taught to believe about nature. A question formed in Bell's mind: was this bizarre nonlocality just a peculiarity of pilot-wave theory, which might be avoided in some better theory yet to be developed? Bell's theorem provides an answer: no, nonlocality is a feature of nature, regardless of what theory we might come up with.

In a nutshell, Bell's argument runs like this. He begins by assuming locality, and from there he derives a contradiction with the statistical predictions of quantum mechanics. Since those predictions have been confirmed by numerous experiments, it follows that locality fails. Nature is nonlocal.

Bell's argument comes in two parts. The first part is a rerun of the EPR argument. If we assume the world is local (no action at a distance), then by considering entangled pairs of electrons, in the singlet state, we can deduce the existence of hidden variables that determine individual outcomes. Now comes the second part. Bell considers measurements made along different directions on each side—so the two magnetic fields are not aligned (Figure 47). The up or down outcomes can be labelled $+1$ and -1. As we have seen, when the magnetic field directions happen to be aligned the outcomes are always opposite: $+1$ and -1, or -1 and $+1$. But when the directions are not aligned, sometimes the outcomes are opposite and

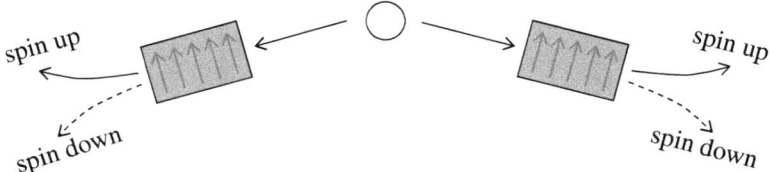

Figure 47. Spin measurements for an entangled electron pair, with magnetic fields pointing along different axes. In the singlet state the outcomes are *not* always opposite. In some cases we find 'up and up' or 'down and down' (as shown).

sometimes they are not: we can also get +1 and +1, or −1 and −1, in some cases. To quantify this we can look at what is called the 'correlation function', which tells us the tendency for the outcomes to be related statistically.[31] For example, in the aligned case where the outcomes always come out opposite, their product is always $(+1) \times (-1) = -1$ and we can say that the outcomes are perfectly 'anti-correlated'. For this case the correlation function is equal to −1, which is a mathematical way of saying that the particles always give opposite outcomes. Now, by imposing the condition that an outcome here is not affected by any measurement done over there, Bell was able to find a general mathematical restriction on the correlation function, which is valid even when the magnetic fields are not aligned. This is the famous 'Bell inequality'. What does it tell us?

Bell's inequality is especially easy to describe when the magnetic field directions differ only slightly, so the angle between them is small. As the angle increases a bit from zero, the correlation becomes bigger than −1, which means the anti-correlation is getting weaker (the outcomes are not always opposite). This is what we might expect. But Bell was able to be more precise: at small angles the correlation function cannot drop below the minimum V-shape curve shown in Figure 48. This is Bell's inequality at small angles. It must be true *if* there is no action at a distance.

Now we have to compare this with the prediction of quantum mechanics. If we apply Born's formula to calculate the correlation function, at small angles we find a parabolic curve, also shown in Figure 48. The quantum prediction, which agrees with the experiments, lies *below* the minimum predicted by Bell's inequality. In other words, at small angles, the correlation stays closer to −1 than locality allows. Somehow, as the

---

[31] Technically, the correlation function is the average of the product of the outcomes.

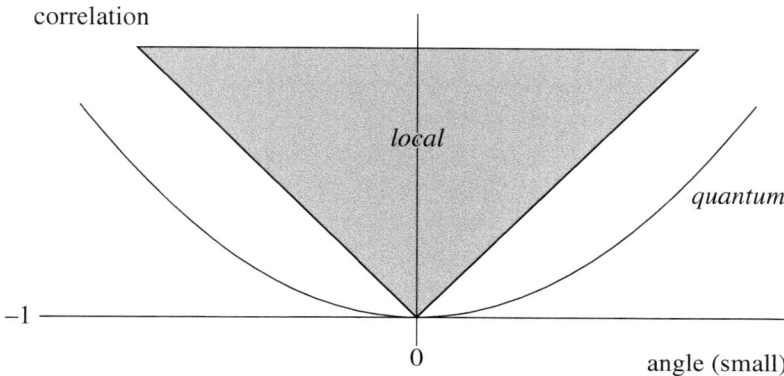

Figure 48. Bell's inequality at small angles. Local theories are confined to the shaded region, a restriction that is violated by quantum theory and by many experiments.

angle is increased from zero, the particles continue to 'do the opposite' more often than they should. It is as if the particles are somehow communicating—instantaneously, even when they are far apart.

This is the scientific evidence for nonlocality. Violations of Bell's inequality—as illustrated in Figure 48 for small angles—have been extensively reported in many laboratory experiments. Entangled particles show a stronger (anti-)correlation than is possible in a local universe. Spooky action at a distance is real.

The earliest violation of Bell's inequality was observed in 1972 in experiments conducted by the American physicists John Clauser and Stuart Freedman. These experiments measured photon 'polarisations' instead of electron spins. For entangled pairs of photons, the same arguments and conclusions apply. Further notable experiments were carried out in the early 1980s led by French physicist Alain Aspect. For this work, Clauser and Aspect shared the 2022 Nobel Prize in Physics with Austrian physicist Anton Zeilinger, who in the late 1990s also carried out significant experiments with entangled photons.

And yet, in the early 1970s, there was widespread opposition to such work. Clauser recalls being summarily dismissed by Feynman who, in an interview in his office, berated the young researcher for wasting time with such pointless and futile experiments. Most physicists believed that in his later years Einstein had become a reactionary, unable to keep up with the quantum revolution. The EPR argument, and its further development by Bell, was not important and best ignored. We now know that, on the contrary, Bell's theorem is one of the most profound results in physics, and Clauser's early experiments rank among the most important scientific experiments of the twentieth century.

At this point we must firmly counter the entirely false claim—which used to be commonplace, and is still heard in some quarters—that the observed violations of Bell's inequality disprove hidden-variables theories (in which unknown hidden parameters determine the apparently random outcomes of quantum measurements). In fact, the observed violations disprove only *local* hidden-variables theories—that is, theories in which outcomes here do not depend on which measurements are done far away. A nonlocal hidden-variables theory, such as pilot-wave theory, is perfectly consistent with the experiments. In such a theory, an outcome here can depend (instantaneously) on which measurements are done far away.

To avoid misunderstandings we should also emphasise that, in these experiments, if we measure outcomes only on one side we will not notice any influence from the other side. It is only when we compare results gathered from both sides that we notice an unusually strong (anti-)correlation—as if the two sides were somehow communicating and working in tandem. So there is no actual signal, only a strange statistical correlation or connection. And we can explain that correlation only if we accept that nonlocality is at work.

While the evidence for nonlocality is strong, still there are caveats. As usual in science, there are implicit or explicit assumptions that can be questioned. There is a huge literature analysing the assumptions behind Bell's theorem. Here we consider what are widely regarded as the three main loopholes.

Firstly, when we measure the spin of an electron, we assume that only one outcome occurs—either it comes out up or it comes out down. This may seem entirely obvious. But it is denied in the many-worlds interpretation of quantum mechanics, according to which *both* outcomes occur—albeit in parallel universes. In one universe the spin comes out up; in another universe it comes out down. We have argued that the many-worlds interpretation would not be necessary if Schrödinger's misstep (removing the trajectories from de Broglie's theory) had not artificially created the notorious measurement problem. Even so, this is a genuine loophole: if there are parallel universes, Bell's argument no longer makes sense, and we cannot conclude that locality is violated. Today, a significant number of physicists find parallel universes more plausible than nonlocality.

Secondly, Bell's argument assumes that there are no physical influences acting backwards in time. Specifically, the initial hidden variables are not affected by measurements performed at later times. Again, this assumption may seem entirely reasonable. In all of our experience, cause and effect act from the past to the future, not the other way around. But some physicists find back-in-time causation more plausible than nonlocality. They try to construct theories in which measurements performed now can have an influence on hidden variables at earlier times. In such theories, quantum correlations between entangled particles can be explained without any need for nonlocality. Only a handful of physicists have worked on this idea, however, which perhaps partially explains why only limited progress has been made with it.

Thirdly, and finally, another possible loophole has an even smaller following. It is a bit tricky to explain. Bell's argument also assumes that the hidden variables do not affect the experimenter's choice of magnetic field directions (that is, the choice of directions along which to measure the spins). That choice will presumably be determined by other things—processes within the experimenter's brain, for example, as the experimenter chooses how to orient the magnetic fields in the lab. But the hidden variables that determine the measurement outcomes are assumed to have no effect on that choice.[32] This assumption allowed Bell to consider different measurement directions, with the same underlying distribution of hidden variables, and so derive his inequality. Some people question this. Maybe the same hidden variables determine both the outcomes *and* the chosen measurement directions?[33] In that case, there is no Bell inequality and no need for nonlocality. The observed correlations between spin measurements here and there can then be explained with a *local* hidden-variables theory. This idea is called 'superdeterminism'. Most workers, including Bell, have regarded this kind of theory as logically possible but scientifically implausible, because it amounts to a vast conspiracy across the whole of nature. To see this, imagine that the experimenter chooses

---

[32] Technically, and more generally, the hidden variables are assumed to be statistically independent of the chosen magnetic field directions.

[33] More generally, perhaps the hidden variables are correlated with the magnetic field directions.

the directions of the two magnetic fields according to the current directions of the wind in two different cities. In a superdeterministic theory the same hidden variables that determine the outcomes of spin measurements on a given pair of electrons somehow *also* determine (or more generally, are correlated with) the direction of the wind. The magnitude of the conspiracy can be multiplied ad infinitum, for example by choosing the magnetic field directions according to movements in the stock market, or according to frequencies of photons received from distant galaxies. Everything, everywhere, would have to be connected in deep ways, just to explain the observed correlations between entangled pairs of electrons.[34] Most physicists regard such theories as scientifically unattractive, and even unsound, though a few still consider them more plausible than nonlocality.

What should we believe? If we discount parallel universes or influences acting backwards in time, and if we disregard the conspiracies of superdeterminism, we seem driven to conclude that nonlocality is a fact of nature. Entangled particles can influence one another instantaneously, no matter how far apart they may be. As Bell put it, in his ground-breaking paper of 1964:

> . . . there must be a mechanism whereby the setting of one measuring device can influence the reading of another instrument, however remote. Moreover, the signal involved must propagate instantaneously . . ..

The instantaneous influence between entangled particles does not diminish with distance, and so is much more radical than the gravitational action at a distance envisaged by Newton. Unsurprisingly, since 1964 many physicists have looked for ways to deny that nonlocality is real. It seems too shocking by far; there must be a way around it. In this book we take a different view: we say it is time to bite the bullet, and to make sense of nonlocality as a fact of nature.

As we have emphasised, in experiments with entangled particles, the results point to nonlocality only when we compare measurements made on both sides, where the joint results show strange statistical correlations. If we confine ourselves to observing only one side, we notice nothing unusual. In particular, there will be no discernible instantaneous signal

---

[34] In pilot-wave theory, in contrast, only entangled particles are deeply connected. Particles that are not entangled can be treated separately.

from the other side. We can deduce the existence of an underlying action at a distance, operating behind the scenes, but we can never control it or use it for signalling. Historically this had led to endless doubts. Is nonlocality real, or is it merely some sort of statistical illusion? For as long as we cannot use it to send practical superluminal signals, there will always be some wiggle room, some reason to doubt it really exists. And so the subject has not moved significantly forward since 1964. We seem to be stuck—caught between knowing on the one hand that locality fails, while on the other hand being unable to grasp nonlocality directly.

This book argues that, to break the deadlock, we must recognise that we are unable to send superluminal signals not because of any fundamental limitation mandated by the laws of physics, but only because we happen to be confined to a state of quantum death. In this state, all things are pervaded by a universal quantum noise described by Born's formula—a sort of quantum fog that prevents us from seeing the quantum world clearly, and which prevents us from controlling nonlocality and using it for practical signalling. To send instantaneous signals, we will have to escape from quantum death and find particles that violate Born's formula. That will entail the discovery of a radically new physics beyond what is presently known. How to do that will be one of the main themes of the rest of this book. But first, we need to understand how this new physics will change our view of space and time, and in particular how and why Einstein's theory of relativity will have to be abandoned.

4

Transcending Relativity

The end of peaceful coexistence

We have said that the two pillars of twentieth-century physics—relativity and quantum mechanics—are deeply opposed to each other. The first forbids superluminal signalling, while the second implies instantaneous or nonlocal connections between widely separated entangled particles. And yet, the two theories manage to live side by side in an uneasy 'peaceful coexistence'. Quantum nonlocality is widely regarded as a fact of nature, but it cannot be employed for practical superluminal communication. It is as if something is going on behind the scenes, something that should not be allowed by the theory of relativity, and which remains hidden from direct view. Relativity is violated in spirit, but not in practice.

We have also said that this peaceful coexistence can continue only for as long as we are confined to the state of quantum death, where an all-pervasive statistical noise prevents us from properly controlling the quantum world. Born's formula and the uncertainty principle are not laws of physics but peculiarities of quantum death. Should we find a way to escape from this state, we would be able to control the instantaneous connections between remote objects and manipulate them to send super-luminal messages—in blatant violation of the theory of relativity. The peaceful coexistence at the heart of twentieth-century physics would be at an end. We would need to confront superluminal signalling head on, and revise some of our most basic assumptions about space, time, and relativity.

According to relativity, time is different for differently moving observers. This makes superluminal signalling paradoxical, because if one observer sees a signal travelling faster than light, then a moving observer may see the same signal travelling 'backwards in time'. This is a simple consequence of the equations of relativity (as we will explain). For this reason most physicists believe that superluminal signalling must be

Beyond the Quantum. Antony Valentini, Oxford University Press. © Antony Valentini (2025). DOI: 10.1093/oso/9780198853749.003.0005

impossible. But in fact we can make perfect sense of superluminal signals if we abandon the relativity of time and accept that there is one true time throughout the universe. Relativity remains valid at the level of quantum mechanics, but is violated at a deeper level that we are currently unable to see or control. This will be explained in this chapter.

The idea of abandoning the theory of relativity makes some physicists nervous. Relativity is after all a cornerstone of modern physics. Would it not be retrograde to try to undo the relativity revolution? And yet, like any scientific theory, relativity was born in response to experimental puzzles of its time. As we learn more about nature, we should be ready to revise our ideas and our theories. So let us now take a look at relativity, and try to understand why it is so at odds with nonlocality, and how the conflict can be resolved by abandoning the relativity of time.[35]

Relativity in a nutshell

Einstein developed the theory of relativity over a century ago. The electro-magnetic field had recently been understood and physicists were learning how to control it in the lab. Around 1887 the German physicist Heinrich Hertz demonstrated how to 'shake' the electromagnetic field here and send a signal over there—at the speed of light. He set up an oscillating electric current here and detected an electric spark across the ends of a loop of wire over there. What we now call a radio wave had travelled a short distance in the lab—carrying energy across seemingly empty space. The essential technology needed for radio communications was born. By 1902 the daring Italian inventor Guglielmo Marconi was sending reliable radio signals across the Atlantic Ocean.

But there were puzzles about how to reconcile the electromagnetic field with Galileo's relativity of (uniform) motion. In his great dialogue of 1632, Galileo had understood that if we are on a (steadily) moving ship and unable to look outside, we cannot tell if we are moving or not. For example, if we drop a ball it falls straight down to the floor, just as it would if the ship were at rest. Today we would say that the laws of physics are the same on the moving ship as they are at rest on dry land. This means that uniform motion is 'relative'—we cannot say who is truly moving and who is not. The discovery of the electromagnetic field seemed to fly in the face

---

[35] The rest of this chapter is not needed to follow the main thread of this book.

of all this. People thought of the field as something like a fluid or medium filling space, which they called the ether. Surely, either we are moving through the ether or we are not. The ether seemed to define a true state of rest—what physicists call a 'preferred rest frame'. Experiments were done to detect the earth's motion through the ether, but came up with nothing. Surely the moving earth was not dragging the ether along with it? People were puzzled.

In 1905 Einstein famously resolved the conundrum by deploying light signals—or electromagnetic waves—to redefine time for moving observers. If this was done in the right way, moving observers would see the same physics as observers who are at rest. We could then safely declare that, after all, uniform motion *is* still relative—we just have to accept that time is different for differently moving observers. Time itself is not absolute but relative.

So, for example, an experimenter in the lab has clocks to measure time, and an experimenter passing by in a spaceship also has clocks to measure time. But in general the clocks in the lab and the clocks in the spaceship will read different times. The faster the spaceship streaks past the lab, the bigger the difference. At ordinary speeds the difference is tiny, but for speeds close to that of light the difference becomes large. The difference can be calculated with certain formulas called 'the Lorentz transformation'—named after the Dutch physicist Hendrik Lorentz.[36] Why are these formulas named after Lorentz and not Einstein? Because Lorentz worked them out before Einstein did. But Lorentz's *interpretation* of the formulas was different from Einstein's—an important point we will come back to.

The conflict with nonlocality

With that thumbnail sketch of relativity, let us now try to understand the conflict with nonlocality. According to the relativity textbooks, if we could send an instantaneous signal from one end of our lab to the other, then a moving experimenter—passing by in a spaceship—would see the signal going back in time (Figure 49). This follows from Lorentz's formulas. We will explain why later, but for now let us simply accept that it follows from the formulas. As seen from the spaceship our signal goes back in time. But

---

[36] The Lorentz transformation also relates distances as well as times (for different observers), but this does not matter here.

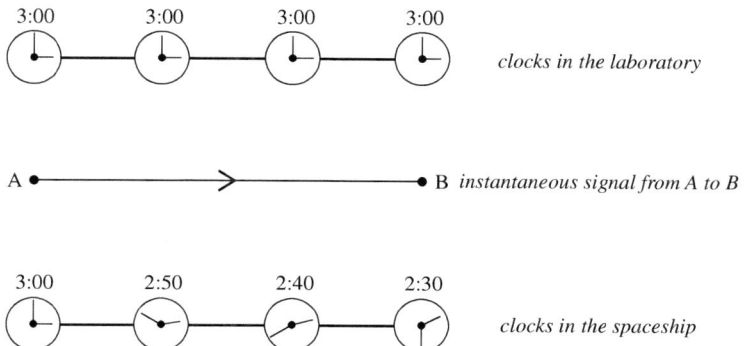

Figure 49.   An instantaneous signal in the lab (clocks above) appears to travel backwards in time in the spaceship (clocks below).

that seems absurd. Hence the common conclusion: instantaneous signalling must be impossible. Most physicists agree with this. However, in a subtle way, the argument is misleading.

To get a grip on this, we have to tread carefully and think in concrete terms. Here is a question to focus on: what exactly is *meant* by the phrase 'going back in time'?

Take this example. Pauline catches a plane in Paris at 3:00 p.m. local time—as measured by clocks in Paris. She lands in London at 2:30 p.m. local time—as measured by clocks in London. Has she gone back in time? We would not normally say so. We know that Paris and London are in different time zones. Because Paris is further east, and because of the eastwards rotation of the earth, in Paris the Sun rises before it does in London. For this reason clocks in Paris are *set* one hour ahead of clocks in London. So when Pauline's plane took off at 3:00 p.m. in Paris, it was only 2:00 p.m. in London. Her flight lasted half an hour—landing in London at 2:30 p.m. But it would be misguided to say that she arrived half an hour before she left. In reality, she arrived half an hour after she left. What really happened is that, before she took off, clocks in London were already set behind clocks in Paris, making it look *as if* she travelled back in time (Figure 50).

With this in mind let us reconsider our problem—that if we send an instantaneous signal in the lab, then an observer in the spaceship sees our signal 'going back in time'. What we really mean is that, as judged by clocks in the spaceship, the signal appears to arrive before it was sent

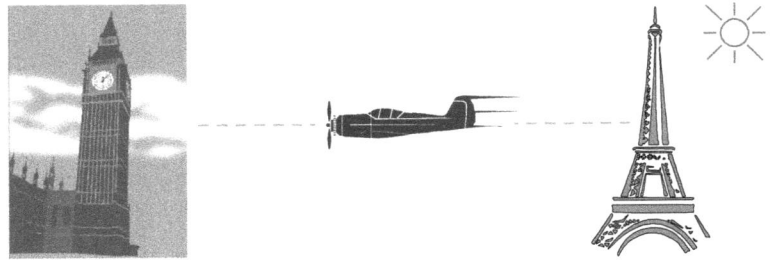

Figure 50. Travelling back in time? Clocks in London are set one hour behind clocks in Paris. A flight from Paris to London arrives 'before' it leaves.

(Figure 49). At first sight that sounds absurd. But we just saw how Pauline can arrive in London 'before' she left Paris—and we understood that this is merely a peculiarity of the way the clocks were set in the two locations. Might something similar be going on with the clocks in the moving spaceship? To find out, we have to ask how those clocks were set. In particular, how exactly were the clocks set to read the same time all across the spaceship?

How do we synchronise our clocks?

We need to consider how to synchronise clocks—that is, how to set them to read the same time. Normally this is so easy we do not even think about it. If you want your clock to read the same time as mine, you simply look at my clock and set yours to read the same time. What is the problem? Well, there is no problem if the clocks are right next to each other. But what if they are far apart? If you look at my clock from far away, it takes time for the light to reach you. So you will really see my clock as it was earlier. To know how much earlier, you need to know how far away my clock was and how fast light travels. In everyday life these effects are so small we can ignore them, but over large distances these effects can become important. For example: if light travels at 186,000 miles per second, and if I am 186,000 miles away from you, then you are seeing my clock as it was one second ago. So if you see my clock reading one second before midday, you understand that for me it is already midday and you need to set your clock accordingly. Fair enough, we can take this into account whenever we synchronise our clocks. But now we run into a problem—or at least,

Einstein ran into this problem when he first thought about it. We have said that, to take into account the time lag when we look at a faraway clock, we need to know how fast light travels. And we have said that in the lab light travels at 186,000 miles per second. But what about in the spaceship? Is the speed of light in the spaceship the same or different?

Before Einstein, physicists assumed that if we ran towards a light wave it would approach us faster, and that if we ran away from it the wave would approach us more slowly. This is certainly true for water waves or sound waves: the relative or effective speed of the wave depends on how fast the observer is moving. We can, for example, easily catch up with a water wave. Think of a surfer riding an ocean wave: with respect to the surfboard the wave is not moving at all, the relative speed is zero. Similarly, we might expect that if we travelled fast enough we could catch up with a light beam. But light waves are strange. As we have said, light waves were once thought to be ripples in the ether. People tried to measure the speed of the earth as it moved through the ether—like measuring the speed of a ship as it moves through the ocean—but came up with nothing. To explain this, in 1905 Einstein made a radical claim: *the speed of light is always the same*, no matter how fast the observer is moving. Head towards it, or head away from it, the speed of a light pulse will always appear exactly the same. This means that, try as we might, we can never catch up with a light beam: however fast we go, it always moves away from us at the same speed. That is quite mind-bending: it is like saying that, no matter how fast a ship sails, an ocean wave will always move past it at the same speed. But this claim, strange as it may seem, allowed Einstein to restore Galileo's relativity and explain why no one could detect the earth's motion through the ether.[37]

Let us now return to our problem of how to synchronise clocks in the spaceship as it streaks past the lab at high speed. Now it is easy. According to Einstein, within the spaceship light still travels at the same speed—186,000 miles per second. So as we look at faraway clocks within the spaceship, we can set them all to read the same time, just as we do in the lab, by taking into account how long it takes for the light to reach us. Does that sound right? Einstein certainly thought so. But there is a subtlety here, and we need to think it through carefully.

---

[37] More formally, the claim that the speed of light is the same, no matter how fast the observer is moving, follows from the assumed relativity of uniform motion.

laboratory

Figure 51. Synchronising clocks in the lab. The flash of light reaches the clocks at the same time.

It is helpful to look at an example. Suppose we want to synchronise two clocks at opposite ends of the lab (Figure 51). To do this we fix a light bulb exactly midway between the clocks. The bulb is switched off. We have three assistants: one near the clock on the left, one near the clock on the right, and one with a finger on the light switch. The last assistant flicks the switch and turns on the light. When the assistant on the left sees the light flash they immediately set their clock to 3:00 p.m. When the assistant on the right sees the light flash they also immediately set their clock to 3:00 p.m. Because the light bulb is midway between the two clocks, and because light moves at the same speed to the left and to the right, the assistants standing by their clocks will see the light flash at the same time, and so they will set their clocks to read 3:00 p.m. *at the same time*. The clocks are synchronised. As the clocks continue ticking, they will continue to read the same time. Easy enough.

Now suppose we want to synchronise two clocks at opposite ends of the spaceship. Einstein tells us that in the spaceship the physics is exactly the same as in the lab. In particular, the speed of light is the same. And it does not matter if the light is moving to the left or to the right, the speed is the same. So experimenters in the spaceship can synchronise their clocks by the same method we used in the lab (Figure 52). Again the light bulb is fixed midway between the clocks. The light moves to the left and to the right at the same speed—186,000 miles per second. When the assistants see the light flash they set their clocks to 3:00 p.m. Again the clocks are synchronised. So far, so good?

Now for the tricky part. While the people in the spaceship are busy synchronising their clocks, let us take a look at what they are doing from the point of view of people in the lab (Figure 53). Let us think about the

spaceship

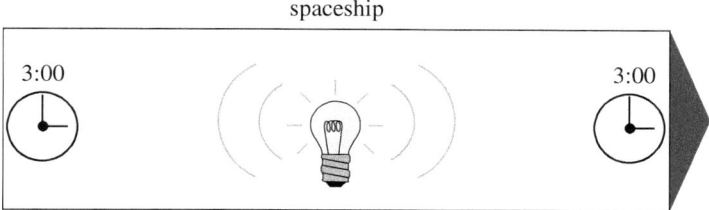

Figure 52. Synchronising clocks in the spaceship. The flash of light reaches the clocks at the same time.

moment when the light bulb is switched on. A light pulse starts to move to the left, at 186,000 miles per second (as measured in the lab, remember the light speed is always the same). But now the clock and the assistant—at the left-hand end of the spaceship—are also moving towards the pulse. So it will take less time for the light flash to reach the assistant (or, if we prefer, for the assistant to reach the light flash). Similarly, when the bulb is switched on a light pulse also starts to move to the right, again at 186,000 miles per second. And now the clock and the assistant—at the right-hand end of the spaceship—are moving away from the pulse. So now it will take more time for the light flash to reach the assistant. So what happens? The upshot is clear: the assistant on the left sees the light flash before the assistant on the right sees the light flash. This means: the assistant on the left sets their clock to 3:00 p.m. *before* the assistant on the right has even seen the flash. By the time the assistant on the right sees the flash and sets their clock to 3:00 p.m., the clock on the left has already ticked its way *past* 3:00 p.m. And so, when all is said and done, the clock on the right lags behind the clock on the left. The clocks are not synchronised.

If all that sounds awkward, it is really quite simple. We can make the same point with sound waves. Imagine someone shouting 'three o'clock'. If you are running towards that person, the sound wave will reach you a bit earlier. If instead you are running away from that person, the sound wave will reach you a bit later. If you are carrying a clock and you set it to three o'clock at the moment you hear the person shout 'three o'clock', then your clock will be set forwards a bit if you are running towards them, and your clock will be set backwards a bit if you are running away from them. Simple enough.

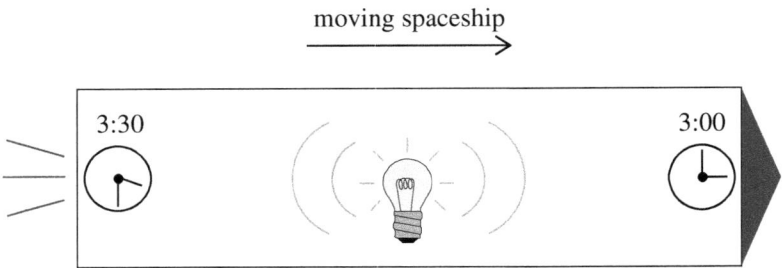

Figure 53. Synchronising clocks in the moving spaceship. As seen by observers in the lab, the flash of light reaches the clock on the left *before* it reaches the clock on the right. As a result the clock on the right lags behind the clock on the left (time difference exaggerated for clarity).

Instantaneous signalling revisited

So now we might be excited. When we sent an instantaneous signal in the lab, our friends in the spaceship said that our signal travelled back in time—which sounded like a real problem. But now it seems clear there is no problem at all. The signal does not travel 'back in time'. What is happening is very simple: the clocks in the spaceship have been *set* so that clocks on the right lag behind clocks on the left. And so—obviously— a signal sent instantaneously from left to right *appears* to go back in time (Figure 49). But this is no more mysterious than Pauline catching a plane in Paris and 'going back in time' to London (Figure 50). The clocks have been set to make it appear that way. There is nothing wrong with instantaneous signalling after all.

At this point your friend who studied physics at university will object: who are the people in the lab to say that the clocks in the spaceship are not properly synchronised? The whole point of Einstein's relativity is that all (uniformly moving) observers are equally valid. We could just as well ask the people in the spaceship to watch what goes on in the lab and *they* will say that the *lab's* clocks are wrongly synchronised. The argument works both ways. So who is right?

According to Einstein, there is no right or wrong on this one. 'Simultaneity' is relative, not absolute. In other words: if two events happen at the same time in the lab, they do not happen at the same time in the spaceship— and vice versa. Time is different for different observers. There is no point arguing about who is right (or wrong). No one wins the argument. We simply have to accept that there are different but equally valid descriptions of what is happening. This is relativity, after all. Counterintuitive as it may seem, relativity works—but only for as long as there is no superluminal signalling.

If we do allow instantaneous signalling, however, relativity runs into trouble. One observer might see the signal travel instantly across space, but another observer may see the same signal travelling back in time. According to relativity, both points of view are equally valid. We cannot discount the second view as an illusion caused by a misleading setting of clocks. We then have to accept that signals *really can* go back in time. This means that, in principle, you might send a signal into the past instructing someone to shoot you dead if you so much as pick up this book—raising the question of how it is that you are in fact now reading it. While some physicists try to make sense of back-in-time signals, most regard them as paradoxical and therefore impossible. In this book we take it for granted that we have to avoid signalling into the past. But how?

Back to Lorentz: or reversing relativity

There is an easy way to do it. We have to accept that the different points of view are *not* equally valid. The people in the spaceship are mistaken: they have set their clocks incorrectly. When they appear to see our instantaneous signal going back in time, that is effectively an optical illusion. And the root of the problem is that their clocks are not properly synchronised. But how do we know it is the people in the spaceship who are mistaken and not the people in the lab? Does the argument not work both ways? The answer is simple: if the people in the lab see the signal travelling instantly from one point in space to another, then their clocks are set correctly. In fact, if we were able to control nonlocality and apply it to send practical instantaneous signals, then it would make sense for us to use *those* signals to synchronise our clocks in the first place—instead of relying on the nineteenth-century technology of electromagnetic waves (such as light or radio waves). The problem of messages apparently going back in time would disappear.

Your friend who studied physics at university might not be happy with this answer. Generations of students have been taught that Einstein's relativity is somehow inevitable, as if it were the only sensible way to think about time. To say that only one observer is right flies in the face of everything they have been told. But in fact this was Lorentz's interpretation of his formulas—only one observer is right. Despite what students are usually taught, Lorentz's interpretation works just as well as Einstein's. The mathematical equations are the same. The observable results are the same. It is only the physical interpretation that is different.

The historical facts are a bit more convoluted, however, than we are saying. Lorentz had the correct transformations for distances and times, but his transformation formulas for charge density and current were not quite right. That was fixed in 1905 by the French physicist and mathematician Henri Poincaré. With that amendment, Lorentz's interpretation is fully equivalent to Einstein's.[38]

Now Lorentz did not know about nonlocality and so had no way of telling *which* observer was right. But he assumed there is a correct observer or 'reference frame' (to use the jargon), with a true time and correctly synchronised clocks. That observer is at rest in the ether. Moving observers can define different times if they wish, and this can be done so that they see—or appear to see—the same laws of physics. But according to Lorentz this is a peculiarity or 'symmetry' of the mathematical equations: if we redefine time for moving observers in an appropriate way, we can make the equations or laws *look* the same for the moving observer. Because the equations or laws look the same, a moving observer is unable to tell that they are really moving. Are we at rest or in uniform motion? We cannot tell the difference. Even so, according to Lorentz, there really is a difference.

The reader might wonder what happens to other well-known aspects of relativity physics, such as moving clocks running more slowly and moving rods contracting. On the Lorentz interpretation these phenomena turn out to be general effects of uniform motion. The equations and predictions are the same as usual.[39]

At this point, enter Einstein. The great man's real innovation was this: if we cannot tell the difference between rest and uniform motion, then there *is* no difference. There is no true state of rest. Lorentz's true time is a fiction. The times defined by different observers are all equally valid. This is the interpretation taught in our universities and in our textbooks.

---

[38] For historical completeness it is also worth noting that, in 1905, Poincaré had the full mathematical formalism of what we now call 'special relativity' (independently of Einstein's work in the same year).

[39] For a clear account, aimed at both students and practising physicists, see Bell's article 'How to teach special relativity', republished in his book of 1987.

But it is worth pausing to ponder. If we cannot tell the difference, then there is no difference. Are we sure about that? We again detect the ghost of the ancient conflict between perception and reality, haunting us down the ages.

The power of Einstein's position lies in the claim that we can *never* tell the difference. We cannot tell the difference between rest and uniform motion—as Galileo famously found in his moving ship—because that is a fundamental law of the universe. Lorentz's approach assumes there is a true state of rest, which by the way can never be known. Why would a scientist waste time even thinking about something that can never be known? For this reason, Einstein's interpretation is widely seen as unassailably superior to Lorentz's.

But now we need to rethink all this. What if we cannot tell the difference between rest and uniform motion, not because of any fundamental law, but because of our present conditions—specifically, that we happen to live in a state of quantum death? Perhaps the 'laws' of relativity are not laws at all, but merely peculiarities of that special state?

We have said that, in our present conditions, an all-pervading quantum noise prevents us from using entangled particles for instantaneous signalling. So we are forced to synchronise our clocks with light signals. Because we never encounter instantaneous signals, we never run into the problem of signals going back in time. We can consistently assume that uniform motion is relative, and that time and simultaneity are relative. Relativity works, but only because we are confined to quantum death. If one day we manage to escape, we will be able to send instantaneous signals—and use them to synchronise our clocks. We may then discover that Lorentz was right after all: there is a true state of rest and a true (absolute) time.

Some readers might wonder how Lorentz's approach can be extended to Einstein's theory of gravity, in which space and time are curved (Chapter 8). The answer is simple enough. Even in strong gravitational fields, it is usually possible to describe physics with a true time.[40] In Einstein's theory there are generally many different ways of defining time. But, as in the Lorentz approach, we can assume that only one choice is correct and that nonlocality acts instantaneously as measured by this 'absolute time'.

---

[40] Technically, this can be done if there exists at least one 'foliation of spacetime by spacelike hypersurfaces'. Many physicists think that physical spacetimes are in fact always of this 'globally hyperbolic' class.

This book argues that quantum physics succeeds only because of our confinement to quantum death. It turns out that the same is likely to be true of relativity physics. If ever we manage to escape quantum death, both theories will break down. The two great physical theories of the twentieth century, relativity and quantum mechanics, stand and fall together.

Relativity and pilot-wave theory

In retrospect, none of this should be too surprising. We saw in Chapter 3 that de Broglie abandoned Newton's first law—according to which a free body moves uniformly in a straight line. But that law is deeply connected with the relativity of uniform motion. If we abandon it, relativity must fall as well. To explain this properly would take us too far afield, but here is the gist of it.

Let us say a tennis ball is lying still on the table in the lab. With respect to the lab, the ball is at rest. Someone passing by in a uniformly moving spaceship looks out of the window and sees the ball moving (with respect to them) at high speed. They may ask themselves: *why* is that ball moving uniformly at high speed? Newton would say that nothing is making it move, because his first law states that uniform motion is natural and does not need to be explained (unlike acceleration, which is caused by forces). This also makes sense from the point of view of relativity. The observer in the lab sees a ball at rest, while the observer in the spaceship sees a ball moving uniformly. Relativity asserts that both viewpoints are equally valid. This implies that a ball can move at uniform speed without a cause. So Newton's first law and the relativity of uniform motion are deeply connected. If we abandon the first law, as de Broglie did, we can expect to have to abandon relativity as well.

We can make the same point in a different way. In de Broglie's theory velocity *does* have a cause. If we remove the cause, there will be no velocity. In other words, the natural state of motion is rest. And so, inevitably, in pilot-wave theory there must be a true rest frame—as assumed by Lorentz.

We see that pilot-wave theory has a powerful internal logic. The non-locality of the theory impels us to reverse Einstein's move and return to Lorentz's original interpretation of his formulas, with a true time and a true state of rest. The dynamics of pilot-wave theory, in which velocities have a cause, drives us to the same conclusion. All the pieces fit, like a jigsaw puzzle.

The internal logic of pilot-wave theory is all the more remarkable for having gone almost entirely unnoticed. Even de Broglie missed the point that his dynamics is conceptually at odds with relativity. For decades many workers in the field have pursued a 'relativistic' pilot-wave theory, in which time and (uniform) motion are supposedly relative. Unsurprisingly, they have not succeeded. At the deepest level of pilot-wave theory, relativity must be abandoned.

Relativity and quantum death

Before moving on, we need to say a bit more about how relativity emerges in the state of quantum death. We have seen that Einstein, together with Lorentz and Poincaré, discovered relativity while trying to understand the electromagnetic field. So we should look more closely at what pilot-wave theory has to say about the electromagnetic field.

The quantum theory of the electromagnetic field was first written down by the British physicist Paul Dirac in 1927. That was the beginning of what we now call 'quantum field theory', which is widely applied to describe quantum physics at high energies. The energy of the field is found to take on discrete values—rather like atomic energy levels. These are often interpreted in terms of particle-like excitations called photons. But as usual the standard theory is quite vague about what is really happening.

The pilot-wave theory of the electromagnetic field was first developed by Bohm in 1952. At first glance it seems different from the kind of pilot-wave theory we have been looking at so far, but really it is based on the same idea. So far we have talked about particles moving in space. What we have now are fields—electric and magnetic fields—which are moving in space and changing with time. But we still have a pilot wave guiding the motion, and acting in a hugely dimensional configuration space that describes the fields. It is not easy to visualise, but mathematically it is a straightforward generalisation of the theory for particles. If we assume Born's formula for the statistics of the field, then as before we find the same predictions as given by quantum theory.

So now what about relativity? The equations Bohm wrote down require that there is a true state of rest. Nonlocality acts instantaneously in this preferred reference frame. But Bohm only considered quantum equilibrium—in other words, the state of quantum death—and so there was no actual instantaneous signalling. In the state of quantum death all the statistics can be calculated from Born's formula. Something remarkable happens when we look closely at those statistics: they satisfy the symmetry originally described by Lorentz. We can define a new time for moving observers, and we find that the equations or laws stay the same. In the equilibrium state of quantum death, we cannot tell the difference between rest and uniform motion. Physicists call this symmetry 'Lorentz invariance'—which, again, is named after Lorentz (and not Einstein) because he was really the first to work it out. The same symmetry turns up in pilot-wave theory—*if* we are confined to the state of quantum death.

But we can go further and ask what happens to the electromagnetic field if we are outside the restricted state of quantum death. We find that Lorentz invariance is 'broken': even if we define a new time for moving observers, the equations or laws are no longer the same and we *can* tell the difference between rest and uniform motion.

Why is Lorentz invariance broken when we leave the state of quantum death? Here is a rough way to see it. Quantum field theory teaches us that so-called empty space is never really empty. What we usually call the vacuum contains small electric and magnetic fields, as well as other fields. The effects of these 'vacuum field fluctuations' have been observed in the form of small shifts in atomic energy levels. They are real. In effect, the vacuum is really an ether. Now here is an interesting idea. We might think that if empty space contains fluctuating fields, then perhaps we can tell whether we are moving (uniformly) by looking at those fields—just as on a boat we can tell whether we are moving by looking at the surface of the water? This is a good question. Pilot-wave theory gives a precise picture of the actual vacuum fields, so we should be able to work out the answer. In the state of quantum death, the vacuum field fluctuations can be calculated from Born's formula and so the answer is the same as in quantum theory: we find that the vacuum is Lorentz invariant, which means that the fluctuations look the same whether we are moving through them or not.[41] So in the state of quantum death, we cannot tell whether we are moving simply by looking at the vacuum fields. But in pilot-wave theory we can go further: we can ask what happens if we are not in the state of quantum death. We find that, in general, the field fluctuations are different from usual—and they do *not* look

---

[41] Technically, the quantum spectrum of vacuum fluctuations depends on the 'Hamiltonian' of the field, and for the Hamiltonians found in nature the spectrum is always Lorentz invariant.

the same whether we are moving through them or not. As a rough analogy with the boat on the water: in the state of quantum death the surface of the water is so smooth we cannot tell if we are moving, while outside the state of quantum death the surface of the water is rough and now we *can* tell if we are moving.

And so Lorentz invariance—one of the cornerstones of modern physics—is merely another peculiarity of quantum death. Outside of quantum death, we can tell whether we are moving (uniformly) by looking at fields in the vacuum. But in the state of quantum death where everything, including the vacuum, is pervaded by quantum noise, we cannot tell whether we are moving. And for as long as we cannot tell the difference between rest and uniform motion, there will always be those who say there *is* no difference. At the end of the day, the only way to settle the argument will be to escape from quantum death and prove there is a difference once and for all.

At this point the reader might be getting impatient. We are by now building up a long list: the uncertainty principle, Born's formula, the impossibility of superluminal signalling, the relativity of time, and now Lorentz invariance. According to most physicists, these are all fundamental laws. But in pilot-wave theory (at least as interpreted in this book) they are no such thing: they are merely peculiarities of quantum death. The most cherished principles of twentieth-century physics turn out to be not laws, but special features of the state we happen to be confined to. Your friend who studied physics at university may be appalled, or thrilled, according to temperament.

While only a small number of physicists are willing to take issue with the uncertainty principle and other basic laws of quantum physics, it is not uncommon for Lorentz invariance to be questioned—at least among workers in high-energy physics. There are several reasons for this. For a start, modern textbooks on quantum field theory avoid Einstein's talk about light signals or clock synchronisation. Those books are trying to describe microscopic fields and particles, not macroscopic light waves or clocks. So instead they treat Lorentz invariance as just another important symmetry of the equations.[42] Put like that, it seems plausible that the

---

[42] Technically, the 'Lagrangian density' is required to be a 'Lorentz scalar'. This modern and direct route to relativity, in fact, has more in common with the approach taken by Poincaré in 1905 than with the 'operational' approach taken by Einstein.

symmetry could break down somewhere—for example, at very high energies. Another reason is that, in some contexts, Lorentz invariance can feel like a theoretical straightjacket. It drastically narrows down the kinds of laws we can even think about, and sometimes it seems worth exploring what might happen if we let go of Lorentz invariance. Some physicists have proposed plausible theories—of particle physics and also of gravity—in which Lorentz invariance is true only at low energies and breaks down at high energies. So far there is no experimental evidence for a breakdown of Lorentz invariance—despite extensive searches—but there seems to be no harm in exploring the possibility.

Attitudes tend to be quite different among physicists working in the foundations of quantum mechanics. There relativity is often treated as something of a dogma. This is ironic because it is precisely in this area that we find the strongest of reasons for questioning relativity: the repeated experimental verification of quantum nonlocality (which we discussed at the end of Chapter 3).

This book argues that, like quantum physics itself, relativity is valid only in a limited regime—in the state of quantum death which we happen to be living in. Relativity works only for as long as we are confined to this state. Should we find a way to escape, quantum physics will fail and relativity will be transcended.

5
Beyond Quantum Physics

Escaping from quantum fog

Today we are trapped in a state of quantum death in which an all-pervading statistical fog obscures our vision of the microworld. The fog is described by Born's formula for quantum probability. Hidden behind the fog there lies a strange universe where everything is connected by invisible instantaneous forces, a sort of multidimensional cobweb that makes every part of the universe instantly aware of what is happening everywhere else. But we are unable to see or control this remarkable reality. The uncertainty principle reigns—not because it is a law of physics, but because it is a peculiarity of the state we happen to be in. That is the central thesis of this book. Which begs the question: will it ever be possible for us to escape the fog and finally see what is really happening behind the scenes?

If we take these ideas seriously, our efforts should be focused on trying to find a way out. That effort begins in the next chapter. First, it will help if we understand quantum death in more detail: what it is and how it came about. We should also take a closer look at the kind of radical new physics that awaits us if we do manage to escape the fog and emerge into a state of quantum life. In this chapter, then, we will try to understand quantum death more deeply—and then dip our toes into the wide open waters that lie beyond quantum mechanics.

The quantum conspiracy—the game is up

We begin by looking more closely at what we call the 'quantum conspiracy'. This is not a confidence trick pulled by nefarious scientists. Nor is it a government cover-up. It is instead a trick played on us by nature. Something is going on behind the scenes, something we cannot control or observe directly. That something breaks the known laws of physics—but is seemingly forever hidden. It is as if nature is playing a game with us. What is this all about?

Beyond the Quantum. Antony Valentini, Oxford University Press. © Antony Valentini (2025).
DOI: 10.1093/oso/9780198853749.003.0006

Let us start with an analogy. Some coins are lying on the table in front of us. Some are showing heads, some are showing tails. Imagine that our friend a thousand miles away claps their hands and instantly—by some magic—all the coins on the table flip over. Imagine further that we are not allowed to track each coin or watch as it flips over. All we can do is *count* the total numbers of heads and tails—before and after. What will we see?

The answer depends on how many coins were showing heads or tails beforehand. Imagine we begin with an even or 50:50 ratio: 50% show heads and 50% show tails (Figure 54). Our friend far away claps their hands and each coin magically flips over. What do we see? Well, if we are only allowed to count the total numbers of heads and tails before and after, then we do not see very much. In fact, we do not notice any change at all. Half of the coins have flipped from heads to tails and half have flipped from tails to heads. The upshot? We still have an even or 50:50 ratio. As far as our counting is concerned, it is as if nothing has happened. The magical coin flipping is real but *hidden*. This might seem like a conspiracy. But in fact it is just a peculiarity of the 50:50 ratio.

To see this let us look at what happens if we start off with an *un*even ratio, say 30% heads and 70% tails (Figure 55). After the magical coin flipping we are left with 70% heads and 30% tails. The ratio of heads to tails has changed from 30:70 to 70:30. Now we *do* see a difference. The magical coin flipping is revealed.

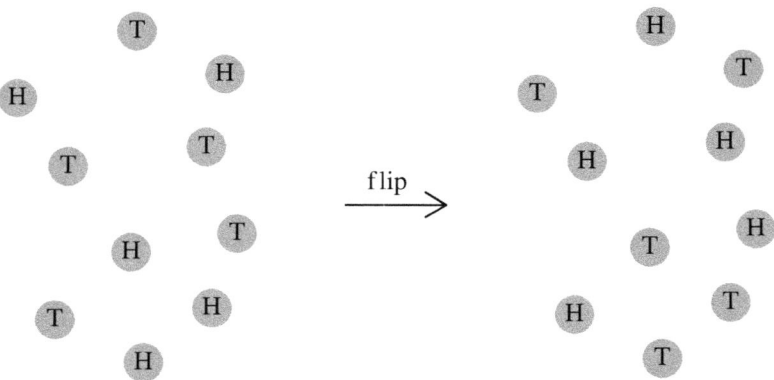

Figure 54. An even ratio of heads and tails remains the same when the coins are flipped. The magic is hidden.

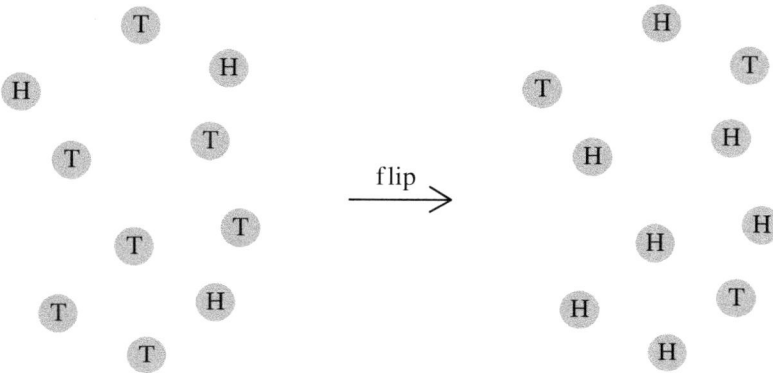

Figure 55. An uneven ratio of heads and tails changes when the coins are flipped. The magic is revealed.

The coin flipping is hidden only if the coins begin with an even ratio. If instead we begin with more tails than heads, or with more heads than tails, there will be a noticeable change. We might put it like this: the magical effect is real for individual coins, but when we count large numbers of coins the effect cancels out (or averages to zero) and becomes invisible—*if* we are in the special state where the coins have an even ratio. If instead the coins have an uneven ratio, the magical effect does not cancel out and remains visible.

This book argues that something similar is going on in the quantum world. A particle here can be instantaneously affected by something happening far away. But in the state of quantum death, for large numbers of particles, such effects cancel out (or average to zero). Nonlocality is hidden and there is no discernible signal. The state of quantum death is fine-tuned, or carefully balanced, so that the underlying nonlocal effects cancel out exactly. This might seem like a conspiracy. But in fact it is just a peculiarity of quantum death. Born's formula for particles is like the 50:50 ratio for coins. Away from that state, the underlying nonlocal effects no longer cancel out. Nonlocality is revealed and there is a measurable superluminal signal.

And so now we can understand what the quantum conspiracy is all about: it is a peculiarity of quantum death and of Born's formula. If we happen to be confined to that state, the nonlocal effects cancel out and the laws of physics *appear* to be conspiratorial. But there is a wider physics in which there is no conspiracy, where Born's formula is violated, and superluminal signalling is possible.

Superluminal signalling

To see all this in detail, let us look again at a pair of entangled particles. We saw in Chapter 2 that applying an electric field to one particle can instantaneously change the direction of motion of the other distant particle. As shown in Figure 56, this is analogous to one of our coins flipping over. If our friend far away applies an electric field to their particle it will 'magically', or nonlocally, deflect the motion of our particle. Our particle can be moving upwards, corresponding to heads, or it can be moving downwards, corresponding to tails. If our friend applies an electric field over there, our 'quantum coin' will instantaneously flip over, from heads to tails or from tails to heads.

Now, if we have an appropriate initial quantum wave, for a large number of entangled pairs 50% of our particles will move upwards (heads) and 50% will move downwards (tails). That is the prediction of Born's formula—and that is what we see in the lab. So what happens when our far-away friend applies an electric field to deflect their particles? The answer is clear. As with the coins, when we look at our particles and count the total numbers moving upwards or downwards, there will be *no change*:

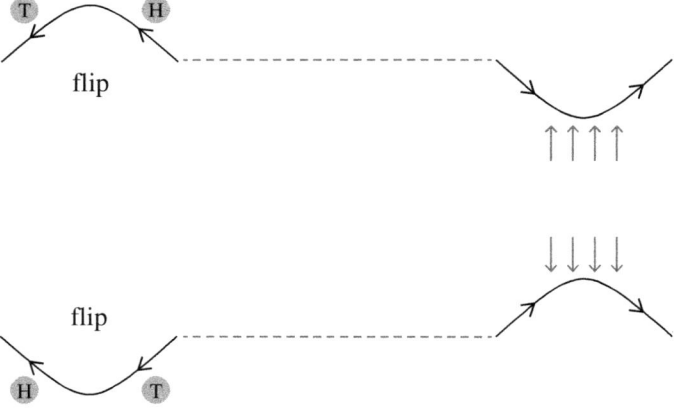

Figure 56. Flipping a quantum coin from far away. If our distant friend applies an electric field to deflect their particle, our particle is also deflected—flipping from heads to tails (above) or from tails to heads (below).

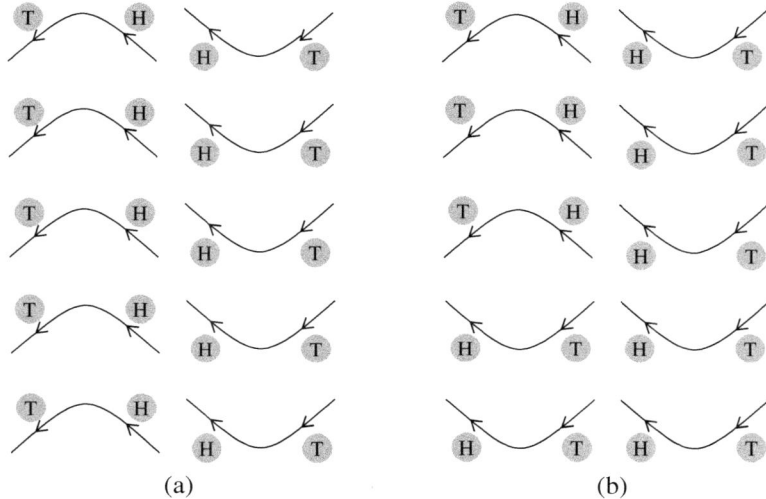

Figure 57. Nonlocality hidden and revealed. In (a), an even ratio of heads and tails remains the same when flipped: the nonlocality is hidden and there is no signal. In (b), an uneven ratio of heads and tails changes when flipped: the nonlocality is revealed and there is a (superluminal) signal.

we are still left with 50% moving upwards and 50% moving downwards (Figure 57a). This means that, even though each of our particles magically and instantaneously reverses its motion in response to our distant friend's electric 'hand clap', when we look at the overall statistics—how many move upwards, how many move downwards—we do not notice any change. The nonlocal flipping is real but hidden, and there is no signal.

Now let us see what happens if, instead, the particles are moving upwards and downwards in a ratio that is *not* even. We should say right away that, for the specified quantum wave in this experiment, an uneven ratio is simply not allowed in quantum mechanics. Born's formula predicts that the ratio will always be even. But in pilot-wave theory we are not necessarily restricted to Born's formula. It is possible—at least in principle—to consider more general ratios. So, for example, let us assume that 30% of our particles are moving upwards and 70% are moving downwards. Clearly, after the nonlocal flipping, we will be left with 70% moving upwards and 30% moving downwards (Figure 57b). The ratio of heads to tails has changed from 30:70 to 70:30 (as in Figure 55). Now we *do* see a difference. The nonlocal flipping is revealed and there is a superluminal signal.

We can draw an important general conclusion, which applies to the real quantum world and not just to our imaginary scenario with coins. The magical—or nonlocal—flipping is hidden only if the particle trajectories begin with the even ratio predicted by Born's formula (Figure 57a). If instead we begin with an uneven ratio, the nonlocal flipping is revealed (Figure 57b).

We might put it like this: the magical nonlocal effects exist for individual particles, but when we count large numbers of particles the effects cancel out (or average to zero)—if we are in the fine-tuned state with an even ratio. That state is precisely what we have called quantum death, where the underlying action at a distance is rendered invisible. If instead we have an uneven ratio, the magical effects remain visible and we can send superluminal signals.

To conclude: if we have entangled particles that violate Born's formula, *we can transmit practical superluminal signals*. This last possibility—not allowed in quantum mechanics but allowed in pilot-wave theory—is a tangible example of what we mean by escaping from quantum death. As we discussed in Chapter 4, this would violate relativity as well as quantum mechanics. We would be able to synchronise distant clocks instantaneously and so define an absolute time across the universe.[43]

There is a more general lesson to be learned here. The fine-tuning we have described occurs not only in pilot-wave theory but in all hidden-variables theories. As we saw at the end of Chapter 3, these are theories in which quantum events are determined by some hidden parameters outside of our control (for example, the exact particle positions in the case of pilot-wave theory). In all such theories it can be shown that, for a certain fine-tuned distribution of hidden variables, the nonlocal effects cancel out exactly and there is no superluminal signalling, while for more general distributions the nonlocal effects are visible and superluminal signalling is possible (as we found in pilot-wave theory).

Quantum relaxation: or how the universe got shaken

All this begs the question: *why* are we currently confined to a state of quantum death described by Born's formula?

---

[43] Broader technological implications of superluminal signalling will be discussed in Chapter 10.

Returning to our analogy with coins, imagine if we lived in a world where coins are always found to show heads and tails in an even 50:50 ratio. For example, we go up into the attic and find an old box full of coins. We open it and count the total numbers of heads and tails. The ratio is even: 50% show heads and 50% show tails. Imagine this *always* happens. We ask a friend to open their purse and empty the coins onto the table and count: again, 50% show heads and 50% show tails. It goes without saying that this is not true in the real world of coins. We might well come across a bunch of coins lying on a table showing more heads than tails for example. But let us imagine a world in which it is true. Could we think of a way to explain it? Here is a simple answer: perhaps all the coins in the world have been *shaken*. Not just once or twice, but for a long time. Think again of that box of coins in the attic. Perhaps last night someone picked it up and shook it up and down for a good while. While the box was shaking up and down the coins inside were flipped around from heads to tails and back and forth. What would we expect to see at the end? An even, or 50:50, ratio of heads and tails. Perhaps our friend's purse was similarly shaken up and down for a good while, with the coins jiggling around inside—again driving the coins to an even ratio of heads and tails. In short: if all the coins in the world were shaken for a long time, we would *expect* that in the end they would always show an even or 50:50 ratio of heads and tails.[44]

So now let us return to quantum mechanics. In the above experiment, with a particular quantum wave, our particles are *always* found to be moving upwards or downwards in an even or 50:50 ratio as required by Born's formula (Figure 57a). It does not matter which particles we use or where they come from. We can perform the experiment with electrons or with photons. We can do it today in London or tomorrow in Paris. We always find the same even ratio. Why? The answer, according to pilot-wave theory, is quite simple: all the particles in our world obey Born's formula because they have all been 'shaken' for a long time.

The particles we find in the lab have not been sitting in a vacuum for billions of years waiting for us to experiment with them. They have a complex history during which time there has been plenty of knocking

---

[44] We are simplifying slightly. The ratio cannot be exactly even if we have an odd number of coins. And there will always be chance 'fluctuations' away from an exact 50:50 mix. But these caveats do not matter when the number of coins is large.

around—plenty of shaking. Take, for instance, an atom of carbon. It first formed a few billion years ago by nuclear reactions inside a star—at temperatures of millions of degrees. Since the formation of the earth that same atom will have spent hundreds of millions of years taking part in all sorts of chemical reactions. When we look at the matter around us we have to recognise its long and complex history. Most of the matter we see was formed inside stars. The simplest atoms, such as hydrogen and helium, formed in the early universe. So we might say that the 'coins' populating our world have all been shaken violently for billions of years. Just as shaking coins drives them to an even or 50:50 ratio of heads and tails, so the violent past history of material particles drives them to the equilibrium state of quantum death described by Born's formula—the dismal state we are now confined to, in which the details of quantum reality are hidden.

This process is called quantum relaxation. We took a quick look at this in Chapter 1, for a simple case of particles trapped inside a box. We talked about how their peculiar wriggling motion drives the particles towards the state of quantum death. It is now time to study this in more detail.

When we know the quantum wave, we can apply Born's formula to calculate the 'quantum probability distribution' (where the particles are likely to be found according to quantum mechanics). Born's formula tells us that the probability of finding a particle near a certain point is given by the squared magnitude (or squared height) of the wave at that point. So the particles are more likely to be found near peaks where the wave is large, and less likely to be found near troughs where the wave is small. Because the quantum wave generally changes with time, the quantum distribution also changes with time. This is shown in Figure 58a for our example of particles inside a box (with 16 energy states in the quantum wave).[45] If we prepare many particles with that same initial quantum wave, and measure their positions at some given time, the distribution always matches Figure 58a. This is what the quantum textbooks tell us. Physics students are told that Born's formula is a fundamental law of physics. No one can explain it, we just have to accept it and move on.

---

[45] For this simple system the quantum distribution is 'periodic in time', returning to its original state after a certain time has elapsed, as can be seen in the figure.

(a)   Born's probability formula

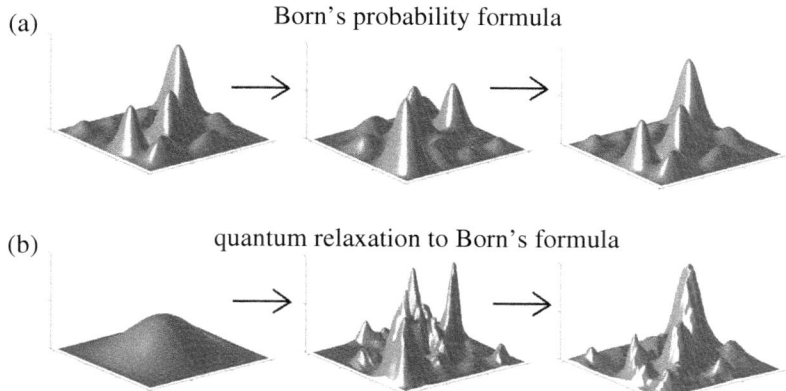

(b)   quantum relaxation to Born's formula

Figure 58. Quantum relaxation for particles trapped inside a box. Row (a) shows the distribution predicted by Born's formula (changing with time and eventually returning to its initial state). Row (b) shows how a more general nonequilibrium distribution can relax to quantum equilibrium, eventually matching Born's formula. (These calculations were performed in 2004 in collaboration with the Swedish physicist Hans Westman.)

But according to pilot-wave theory, Born's formula merely describes the statistics of quantum death. The distribution shown in Figure 58a *can* be explained. And here is how. In pilot-wave theory the particles can start with any distribution, even one quite different from that given by Born's formula. An example is shown in the first frame of Figure 58b. The quantum wave is the same as before, but the particles start off bunched around the middle of the box. At the initial time the distribution of particles does *not* match Born's formula (compare the first frame of the bottom row with the first frame of the top row). What happens then? We can follow the wriggling motion of each particle. And so we can find out what happens. The results are shown in the next two frames of Figure 58b. The distribution sloshes around and ends up looking virtually *the same* as that given by Born's formula (now compare the third frame of the bottom row with the third frame of the top row). Starting from an anomalous state of quantum nonequilibrium, we have already more or less reached the usual state of quantum equilibrium described by quantum mechanics.

What we have just seen is reminiscent of the 'thermal relaxation' studied in introductory physics courses. A simple example is shown in Figure 59. A sample of gas is confined inside a box. The molecules all start in one corner. A physicist would say that the gas is in thermal nonequilibrium. As the

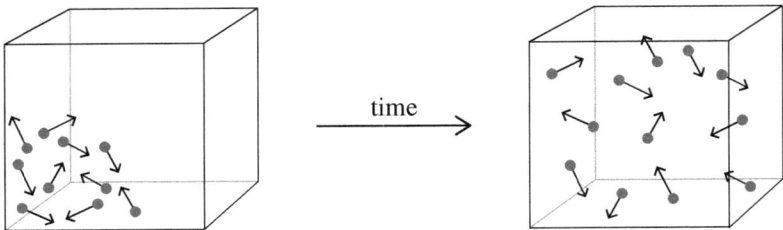

Figure 59. Thermal relaxation for gas molecules trapped inside a box. The molecules move in complicated ways and quickly spread out uniformly over the volume of the container.

molecules move and bounce around they eventually reach a state where the molecules are spread uniformly over the volume of the box. A physicist would say that the gas has relaxed to thermal equilibrium. What we see in Figure 58b is similar: quantum nonequilibrium relaxing to quantum equilibrium.

But the reader should be warned. What we see in Figure 58b is beyond what is even thinkable in standard quantum mechanics. According to Born's formula, for the given quantum wave, the particles must have the distribution shown in Figure 58a. What we see in the first two frames of Figure 58b is a violation of the laws of physics as they are presently understood. According to most physicists, this simply cannot be. And yet, in pilot-wave theory, it is easy to think about. And we can calculate what happens. We find that, despite initially violating Born's formula, the particles end up obeying it anyway (after some time has passed).

In Figure 58b we see an example of the remarkable process whereby our universe may have reached its present equilibrium state of quantum death. *Something like this must have happened to every particle in the universe at some point in its past history.* And so here we are, in a world where everything is infused with the same quantum noise. In our state of quantum death, Born's formula appears to be a fundamental law of physics, because it applies to everything we can see. But according to pilot-wave theory it is nothing of the kind: it merely describes a particular state we happen to be confined to. And we are confined to that state because of past quantum relaxation.

Quantum relaxation explains Born's formula—and, therefore, it also explains what we have called the quantum conspiracy. There *are* superluminal signals in the microworld. But in the equilibrium state of quantum death the signals cancel out, or average to zero, and are not directly noticeable (Figure 57a). Moving away from that state unleashes the underlying nonlocality that makes superluminal signalling possible (Figure 57b). Our universe appears to be fine-tuned. But there is no real conspiracy: we just happen to inhabit a time and place where all particles have reached quantum equilibrium and obey Born's formula.

Subquantum matter

All known forms of matter obey the laws of quantum mechanics without exception. We might call it 'quantum matter'. It is infused with quantum noise, as described by Born's formula. Most physicists believe that all matter is necessarily like this, not only here and now but always and everywhere. Pilot-wave theory says otherwise. In principle we might find matter that violates Born's formula, as shown in the first two frames of Figure 58b. Such matter would violate the laws of quantum mechanics (as well as the laws of relativity). We might call it 'subquantum matter'—a new form of matter radically different from anything seen before. As we will explain in later chapters, subquantum matter might have existed in the early universe, and some of it might have survived to the present day. It might also be created by evaporating black holes. In any case, according to pilot-wave theory, this new form of matter *can exist*, even if we have yet to find it.

Most physicists balk at this. And yet, since the dawn of history, matter has continued to surprise us. The discovery of fire—which must have appeared magical to early humans—was our first experience of controlled chemical transformation. The magnetism of the mineral magnetite fascinated the early Greek philosophers: how could a stone attract a piece of iron with nothing visible in between? An understanding of the electrical and magnetic properties of matter had to wait until the nineteenth and twentieth centuries. The modern theory of chemical elements began with Lavoisier in the eighteenth century and seemed to reach a final formulation by the nineteenth. At the turn of the twentieth century, however, scientists were confronted by an extraordinary violation of what appeared to be the most basic laws of chemistry: matter could disintegrate by radioactive decay, resulting in the 'transmutation' of one chemical element into another (a feat which had been sought after for centuries by alchemists). In the early twentieth century physicists discovered the wavelike properties of matter, again to the amazement of many. Later in the same century the creation and destruction of fundamental particles was observed in high-energy physics. Again and again matter has defied all expectations by the seemingly unending novelty of its forms, properties, and powers. What other surprises might matter have in store for us?

The discovery of subquantum matter would herald a scientific and technological revolution. The first and most obvious novelty would be a breakdown of the quantum conspiracy: the underlying nonlocality of the world would no longer be hidden by quantum fog. When something far away instantaneously affects our particles here, we *will* notice. As we have seen, such effects could be applied to transmit instantaneous signals over large distances. What else could subquantum matter be useful for?

Beating the uncertainty principle

If we had a supply of nonequilibrium or subquantum matter, that reliably violates Born's formula, we could perform many tasks that today's quantum physicists would regard as miraculous. As well as sending instantaneous signals across space, we could also beat the uncertainty principle.

In Chapter 2 we drew an analogy with throwing a tennis ball (Figure 33). To predict where the ball will land we need to know where it is thrown from *and* the velocity at which it is thrown. In classical physics we can (in principle) know these 'initial conditions' as precisely as we like, and so we can predict where the ball will land as precisely as we like. But for quantum particles it is impossible to know all the initial conditions precisely, and so it is impossible to predict where a quantum particle will land in (for example) a two-slit experiment. If we repeatedly fire quantum particles one at a time at a two-slit barrier, in general the particles will start at different points inside the quantum wave and they will land at different points on the backstop (as in Figure 32). There is a scatter of outcomes—which always matches Born's formula. The textbooks say this is inevitable, but pilot-wave theory says otherwise.

According to pilot-wave theory, the scatter at the backstop can be traced to the scatter in the initial positions of the incoming particles—an initial scatter which is not necessary, but merely a feature of quantum death.

The incoming particles already reached quantum equilibrium a long time ago. So when we experiment with them, we always see the statistical spread described by Born's formula. But as we noted in Chapter 2, if we somehow managed to find particles that have not yet reached equilibrium, then firing them at a two-slit barrier would produce strange results. For example, the incoming nonequilibrium particles might all be bunched together in regions much narrower than the quantum wave, in which case parts of the usual wavy pattern at the backstop will be missing (Figure 34). We might say that, if the incoming particles contain less noise than usual, then the scatter of outcomes will be smaller than usual and the standard interference pattern will be disrupted. Such a scenario is unthinkable in quantum mechanics. And we hasten to add that no one has ever seen anything like this in the lab. But according to pilot-wave theory it could happen. The usual uncertainty scatter in the initial conditions could be avoided—for example for relic particles from the early universe (Chapter 7), or for particles radiated by black holes (Chapters 8 and 9).

The uncertainty principle also dictates that we inevitably disturb a quantum particle when we try to measure it. We touched on this in Chapter 1, where we drew an analogy between quantum death and the classical heat death. If our equipment has the same level of jitter or noise as the system we are trying to measure, we will inevitably cause a disturbance. Instead of just seeing the system as it is, we will change it in a way that is uncontrollable. There is no such issue in classical physics (outside of heat death): we can in principle watch a tennis ball without disturbing it. But for quantum particles there is no getting around this: when we attempt to watch a particle, the light we employ (a stream of photons) contains as much quantum noise as the particle does, creating an inevitable and uncontrollable disturbance (Figure 60).

But, again, pilot-wave theory tells us that this is merely a feature of quantum death. If our equipment—including our light source—is in a state of quantum nonequilibrium, it can have *less* jitter or noise than the particle we are trying to measure. We are then able to cause less of a disturbance, and we can see the particle as it really is to greater accuracy. We can finally watch the quantum world without disturbing it. For example, we could track the motion of a particle through a two-slit experiment in real time (Figure 61). The 'hardcore pornography' of quantum physics would no longer be censored.

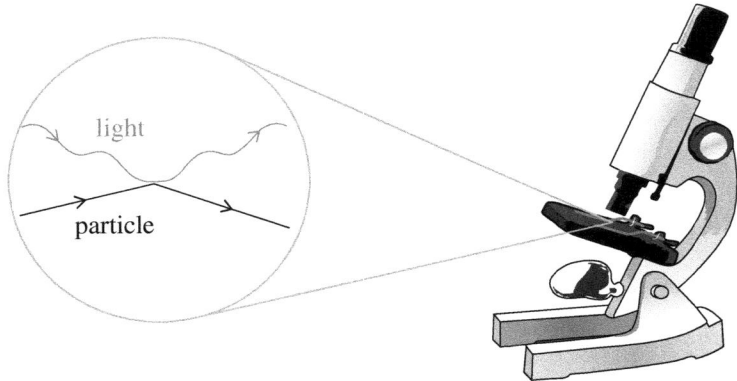

Figure 60. Heisenberg's microscope. We are unable to circumvent quantum noise in the particle we are observing because the light we use to see it contains the same noise (according to quantum mechanics).

All this can be worked out and calculated precisely with the equations of pilot-wave theory. The upshot is that a nonequilibrium light source—or a stream of nonequilibrium photons—can be employed to perform what we call 'subquantum measurements'. These are measurements that allow us to find out what ordinary quantum particles are really doing without causing the usual accompanying collapse of the quantum wave—in complete violation of the central tenets of quantum physics.

And so now we can finally get back to the usual business of theoretical physics, which is to understand and *predict* what nature is going to do. If by a subquantum measurement we observe a particle entering a two-slit barrier at a certain point near one of the holes, and we make this observation without disturbing the guiding wave, we can then apply the equations of pilot-wave theory to predict where the particle will land at the backstop (Figure 61). If such predictions are verified, we will have transcended the limitations of the uncertainty principle.

Subquantum physics

If we could accomplish what we have just described, it would be the end of quantum mechanics and the beginning of a new era. We would be able to probe the quantum world with potentially unlimited precision. We could see our way through quantum fog and keep track of what is really

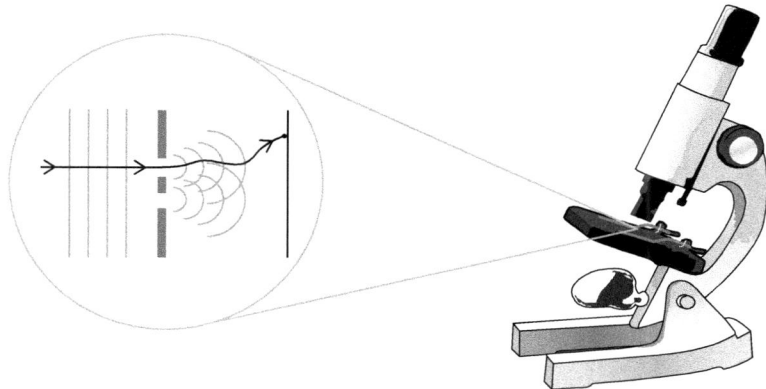

Figure 61.   Watching a quantum particle without disturbing it. If the light source has not relaxed to quantum equilibrium, it can contain less noise than the particle. We can then see what the particle is really doing, even in a two-slit experiment.

happening. We could successfully predict the outcomes of single quantum measurements. But to do all this in practice, we need a source of photons that violates Born's formula. In fact, any kind of particle could be deployed as a probe, as long as it violates Born's formula. In other words, we need a supply of subquantum matter, or subquantum particles.

No such particles are found in the lab today. But, as we will see in Chapters 6 and 7, such particles may have existed in the early universe— and some of them may have survived to the present day. We need to think very carefully about where to find them. Alternatively, it is conceivable that subquantum particles can be *created* by exotic gravitational effects, for example involving black holes. This is discussed in Chapters 8 and 9.

The discovery of particles that violate Born's formula would open up a new world of 'subquantum physics' beyond what is currently known. We would be able to routinely violate what are now regarded as sacrosanct laws of physics. Sending superluminal signals and beating the uncertainty principle would be just the beginning. The past quarter-century has seen a surge in the development of quantum technologies involving the manipulation of information and the construction of new kinds of computers. Such technologies would be transformed (as we will see in Chapter 10). We will have escaped the confines of quantum death and emerged into quantum life.

But all this hinges on us somehow finding or creating subquantum matter. How to do that will be discussed in the next four chapters. We begin by embarking on a journey that will take us to the beginning of time, in a search for signs of what the universe may have been like *before* quantum death.

6
A Message from the Beginning of Time

Looking backwards

Behold the sky at night. We are really looking back in time. The nearest star is about four light years away—it took four years for that twinkling light to reach our eyes. So we are really seeing the star as it was four years ago. For all we know last year it might have exploded in a supernova, but we would not know it for another three years. Looking across our galaxy at the Milky Way, we can see stars thousands of light years away. So we are looking thousands of years back in time. To go back even further we can look at other galaxies. Some are billions of light years away. Our best telescopes are literally looking at the universe as it was billions of years ago. The most distant galaxies found so far are seen by us as they were a few hundred million years after the Big Bang—the gigantic explosion at the beginning of the universe, which took place about 13.8 billion years ago.

We know that in the earliest moments of the Big Bang the universe was extremely hot and dense—and was expanding rapidly. As the universe grew bigger it became cooler and less dense, and the expansion slowed down. Eventually, the tenuous matter started to clump into regions that formed the seeds of galaxies and all the other structure we see in space. A lot else happened besides, of which more later.

But how far back can we actually see? The earliest image of our universe is provided not by starlight but by the cosmic microwave background (Figure 62). This is very faint radiation left over from the Big Bang itself. It is made predominantly of microwaves and pervades the cosmos. It is there in the background wherever we look in the sky. We call it the CMB for short.

Beyond the Quantum. Antony Valentini, Oxford University Press. © Antony Valentini (2025). DOI: 10.1093/oso/9780198853749.003.0007

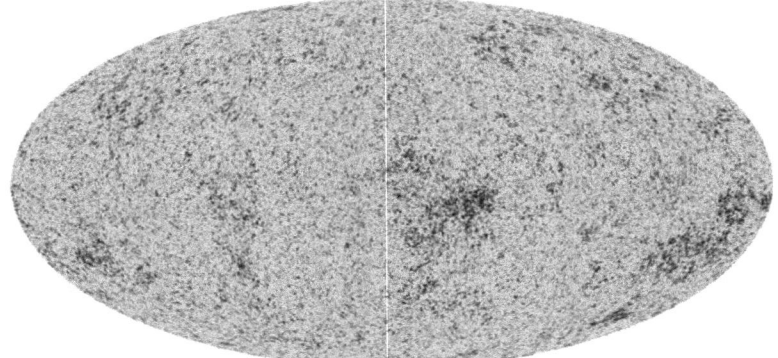

Figure 62. The earliest known photograph of our universe. The cosmic microwave background (CMB) was formed around 380,000 years after the Big Bang. It shows a pattern of hot and cold spots (indicated by light and dark).

The CMB was discovered by accident in 1965 by American astronomers Arno Penzias and Robert Wilson at Bell Labs in New Jersey. They were testing a radio telescope and were puzzled by an unexpected excess noise in the microwave region. At first they thought there must be some stray source. Or perhaps there was something wrong with their equipment. They checked this and that but found nothing. Most puzzling of all, the radiant noise seemed to be coming from all parts of the sky. What was going on? During discussions with Robert Dicke and co-workers from nearby Princeton University, it became clear that they had stumbled on something which had, in fact, already been predicted from Big Bang theory as long ago as 1948 (by American physicists Ralph Alpher and Robert Herman). What they had found was, in effect, a sort of echo of the beginning of the universe—a discovery that revolutionised cosmology (the study of the universe as a whole) and which earned Penzias and Wilson the 1978 Nobel Prize in Physics.

How exactly was the CMB formed? The basic ideas are simple enough. When the early universe was hot and dense it was filled with radiation interacting strongly with a sea of charged particles. As the universe expanded and cooled, the charged particles combined to form atoms of hydrogen. At that point the radiation started to stream freely without touching anything (the atoms are effectively transparent to radiation). And so the CMB was born. This happened about 380,000 years after the Big Bang, when the universe had cooled to a temperature of about 3,000 degrees above absolute zero.[46] So the original temperature of the CMB was pretty hot by terrestrial standards (just below the boiling point of liquid iron). That radiation has been streaming essentially freely through our universe ever since. But in the meantime the universe continued to expand—more slowly, but still expanding. In fact the universe is now about 1,000 times bigger (in every direction) than it was back then. So the wavelength of the radiation has been stretched by about 1,000 times, and so its energy is now about 1,000 times smaller. The upshot: the temperature of the CMB is now about 1,000 times smaller than when it first formed—in other words, instead of 3,000 degrees, it is now only about 3 degrees above absolute zero. Or at least: that is what the theory of the Big Bang tells us.

And that is what Penzias and Wilson found. The excess radiation they had stumbled on had a temperature of about 3 degrees above absolute zero, which was pretty cold by anyone's standards. And it was coming from all parts of the sky, as would be expected if it had a cosmological origin—that is, if it had been created in the early universe.

This was without doubt one of the most dramatic and exciting discoveries in the history of science. Humanity had found the smoking gun of creation. And it was the beginning of a journey which continues to this day, as cosmologists measure the CMB ever more carefully to glean further information about the beginning of time.

We have said that the CMB was formed about 380,000 years after the Big Bang. So when we look at the CMB we are in effect seeing a snapshot of the universe when it was about 380,000 years old. Surely, you might think, we cannot see back any earlier because our line of sight would hit the early and opaque sea of charged particles. In fact, the CMB itself contains detailed information about the universe at much earlier times.

Here is how it works. When we look more closely at the CMB, we find that its temperature is not quite uniform across the sky. There are tiny hot and cold spots as shown in Figure 62. The temperature varies slightly by about one part in a hundred thousand. Why is the temperature of the CMB nearly but not quite uniform? The answer is that the early sea of charged particles was nearly but not quite uniform (not quite homogeneous, as cosmologists like to say). In some places it was slightly more dense than average, in other places slightly less dense. The early universe was not

---

[46] Absolute zero is (approximately) −273 Celsius, so this is $3,000 - 273 = 2,727$ Celsius.

completely uniform—and that is why the radiation left behind is also not completely uniform. Incredible as it may sound, the small wrinkles in temperature we see on the CMB sky are a direct imprint of small wrinkles in the density of matter when the universe was about 380,000 years old. That is certainly impressive. But how does it help us see even further back? The answer is that those small wrinkles grew from even smaller wrinkles that were formed earlier—*much* earlier.

As time goes by and the universe expands, any wrinkles in the density of matter will inevitably grow bigger because of gravitational attraction (a bit more mass here pulls in even more mass here). So the small wrinkles which we see at 380,000 years must have grown from even smaller wrinkles that existed at even earlier times. We cannot see those earlier times directly—but we can apply our understanding of gravity to work out what must have happened.

Today the universe is filled with what cosmologists call 'large-scale structure', by which they essentially mean galaxies and clusters of galaxies. We know that these grew by gravitational attraction. Their formation must have been seeded by small wrinkles (or 'perturbations') in the matter density in the early universe. From our understanding of gravity we can work out how big the early wrinkles in density had to be in order to generate the structure we see today. And once we know how big the early wrinkles in density were, we can work out the size of the corresponding wrinkles in the CMB. The calculation matches what we in fact see in the CMB, proving that we have a good understanding of the growth of structure in the early universe. So far, so good.

But at a certain point the calculations go as far back as we can reliably calculate with classical physics—that is, physics without quantum mechanics. We find very tiny wrinkles (or perturbations) spread over the whole universe, at times just a tiny fraction of a second after the Big Bang, when our entire observable universe was microscopic in size. Cosmologists call these tiny wrinkles 'primordial perturbations'.

We have come a long way since we first gazed in awe at the night sky. But the human mind cannot help asking more questions. Where did the tiny primordial wrinkles come from? Applying only classical physics as updated by Einstein, cosmologists were unable to give an answer. They had to simply postulate a 'primordial spectrum' of wrinkles for which they had no explanation. That was until around 1980, when a radical new theory appeared which was suddenly to transform our understanding of the very early universe.

Quantum mechanics in the sky

Cosmologists were in fact bothered by several gaps in our understanding of big-bang cosmology—puzzles which people had highlighted over several decades but to which no one had a really convincing answer. For example, we have seen that the early universe must have been very smooth (with only very tiny wrinkles). But our theory of the early expansion of space tells us that distant regions of the universe would have had no contact with each other. Why, then, did they look almost the same? Around 1980 this and other puzzles motivated several cosmologists—Alexei Starobinsky and Andrei Linde in Russia, and Alan Guth in the USA—to propose what Guth called the 'inflationary universe': for a tiny fraction of a second after the Big Bang there was a dramatic exponential surge in the expansion of space. This huge and extremely rapid inflation can stretch and smooth out any initial wrinkles, quickly making the early universe almost exactly uniform. But then there is a problem. We are in danger of producing an early universe that is *too* smooth, with no discernible wrinkles at all. There have to be some early wrinkles, to provide the seeds for the later formation of structure. Without them, our universe today would be a featureless blob with no galaxies, no stars, and no planets. So where did the early wrinkles come from? This was answered in 1982 when British theorist Stephen Hawking—as well as Starobinsky, Guth, and others—understood that quantum mechanics was missing from the calculations.

The uncertainty principle adds inevitable quantum wrinkles to the theory. Quantum mechanics deals in probabilities, not certainties. During the early inflating universe, a proper quantum calculation inevitably produces tiny wrinkles in the matter density. And the probabilities for the wrinkles can be calculated with Born's formula. It is found that the quantum wrinkles match pretty well with the primordial spectrum which cosmologists had until then simply postulated. And so the theory of inflation wins twice: not only does it explain why the early universe was so smooth, it *also* explains why it had very tiny wrinkles.

The small temperature wrinkles in the CMB are more easily seen from a satellite in space, so as to avoid the obscuring effects of the earth's atmosphere. In 1992 it was announced that the *Cosmic Background Explorer* or *COBE* satellite had for the first time been able to measure the expected small variations in CMB temperature across the sky. The results were not very accurate but more or less matched what was expected. This generated tremendous excitement. Thus began a new era of 'precision cosmology'. By making more accurate measurements, with new and better satellites, we could finally probe the precise details of what happened during the earliest moments of the Big Bang. The small wrinkles in the CMB may not sound like much, but in fact they contain a huge treasure trove of information about the very early universe.

According to inflationary cosmology, quantum mechanics explains why the CMB sky is not completely smooth. The hot and cold spots seen in Figure 62 are ultimately an imprint of quantum noise. They are a direct result of the uncertainty principle and of Born's formula in action at the very beginning of our universe. And so now it is exciting. By measuring the CMB more accurately, we can test quantum mechanics at the very beginning of time.

It is worth pausing to take in the sheer magnitude of all this. Quantum mechanics with Born's formula is taught to physics students, who apply it to calculate things they measure in the lab—such as the wavy pattern in a two-slit experiment, or the probability for a single atom to emit a photon. Here the theory is being applied to the whole universe, about 13.8 billion years ago, to calculate what we measure today in the sky. This is quantum mechanics on the grandest of scales. And it works—more or less.

'Our most puzzling finding'

After *COBE*, cosmologists focused their attention on a new satellite, the *Wilkinson Microwave Anisotropy Probe*, or *WMAP*, launched by NASA in 2001, which promised to deliver much more accurate measurements of the CMB. The results arrived in several installments, from 2003 to 2013, and confirmed in spectacular detail what *COBE* had found. The CMB temperature wrinkles more or less matched what cosmologists had postulated decades earlier in order to explain galaxy formation. And they more or less matched what was expected from inflationary theory—to the point where people started using the results to probe the finer details of inflation itself. How long did the inflationary expansion last for and what kind of 'field potential' had caused it? These and other technical details of the underlying inflationary model could now to some extent be read from the CMB sky. Some versions of inflation could be ruled out, others could be constrained. We were now truly in the era of precision cosmology and confidence was high. A few voices raised concerns about possible anomalies in the *WMAP* data at large distances in space. But those voices tended to be dismissed.

After all, at larger distances the data were sparser and so more subject to error. There was no need to worry.

In 2009 the European Space Agency launched the *Planck* satellite, named after the German physicist Max Planck who in the year 1900 had in effect started the quantum revolution. *Planck* promised to yield results even more accurate and complete than those of *WMAP*. When the first results came online in 2013 there was a flurry of activity to check how well they agreed with what had already been found by *WMAP*. Two independent experiments, with two independently constructed satellites, broadly confirmed the same picture.

But not entirely. The *Planck* team reported what seemed to be small but significant anomalies at large distances (or at large angles on the sky). There was one anomaly in particular, which the team called 'our most puzzling finding'. On large scales the wrinkles were a bit smaller than expected. Technically, looking at the statistics of hot and cold spots that are widely separated on the sky, the 'power spectrum' was found to be about 10% smaller than expected (Figure 63). Cosmologists call this a power deficit, and the anomaly came to be known as the large-scale power deficit. But was it real?

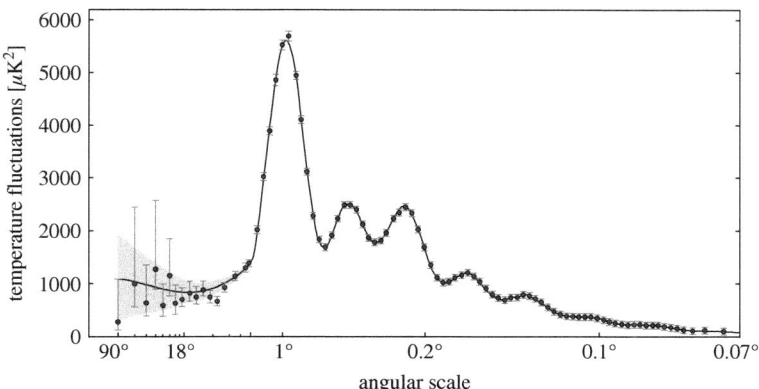

Figure 63. 'Our most puzzling finding'. The statistics of temperature wrinkles on the CMB sky depends on the angle between the hot and cold spots. (Technically, the figure shows a plot of the 'angular power spectrum' as a function of angular separation on the sky.) At small angles—shown towards the right—the measured statistics (black dots) closely match what is expected (solid curve). At large angles—shown towards the left— the measured statistics generally fall about 10% below what is expected (as reported in 2013 by the *Planck* team).

Some cosmologists thought the *Planck* team had overstated the significance of this finding. If we are looking at small distances or small angles, we can divide the sky into a large number of small regions, and so we can gather lots of statistical data. But when we are looking at large distances or large angles, we have to divide the sky into a small number of large regions, and so now we have much less data. And the less data we have, the less accurate our statistical analysis will be. And so there is a fundamental limit on the accuracy of our statistical analysis at large distances or large angles (known as 'cosmic variance'). A reported power deficit at large scales can then be dismissed as a statistical fluke and not a real physical effect.

In everyday life it is sometimes said that there are 'lies, damned lies, and statistics'. This is an exaggeration, of course. Statistical analysis of data plays a crucial role in the modern scientific method. But there are grey areas, and when we are approaching the limits of what can be measured reliably there are sometimes disagreements over how to assess the data— even among the experts. As a case in point, in hindsight we now know that the *WMAP* team had already seen a comparable low-power anomaly but had been less inclined to take it seriously or to highlight it.

But the low-power anomaly is difficult to dismiss because it seems to be accompanied by *other* anomalies at the same large scales. Cosmologists usually assume that the statistics of the CMB will be the same in all directions. The temperature wrinkles vary from one part of the sky to another, but when we look at statistical averages we should find essentially the same pattern wherever we look on the sky. This is known in the jargon as 'statistical isotropy'. And that is what most models of inflation predict: the statistics should be the same in all directions. This basic prediction matches well with what we see. Except at large distances. The *Planck* team reported what they called 'significant' evidence for a violation of statistical isotropy at large scales. What is more, this was found in the *same* large-scale region as the power deficit. One anomaly can be dismissed as a fluke, but two anomalies in the same region may well mean something.

Evidence for this second kind of anomaly had in fact already been brewing for a while in the earlier *WMAP* data. In 2005 Portuguese cosmologist João Magueijo at Imperial College London, with PhD student Kate Land, analysed the *WMAP* data and claimed to find a mysterious alignment at large scales. It was as if the statistics of the wrinkles were lining up along a preferred direction in the universe, which they playfully dubbed the 'axis of evil'. For many cosmologists this was just too weird. Why would the whole universe be aligned along one direction? Again opinion was divided over how significant the evidence really was. But a similar anomaly persists in the *Planck* data. Is nature trying to tell us something?

In their report the *Planck* team suggested that theorists should work harder to produce a model that could explain both kinds of anomalies in one go. After all, seeing as the anomalies were found in the same region, it was reasonable to suppose that they might have the same origin. What might that origin be?

Since 2013 various ideas have been proposed, usually involving more complicated models of inflation. Workers in the field remain sceptical. Cooking up a more complicated model, to explain something that may or may not even exist, somehow fails to convince. Theorists are more impressed when a model *naturally* explains what we see—naturally, in the sense of flowing from the internal logic of the theory, instead of being put in by hand. Until some such explanation comes along, most theorists are likely to continue to ignore the anomalies and hope that they will somehow go away. Perhaps there is some error in the data. After all, measuring the CMB involves extracting a weak primordial signal from lots of background noise (such as emission from dust in our galaxy). Any theorist reading the papers released by the *WMAP* and *Planck* teams is bound to be humbled by the extraordinary attention to detail, the rigorous cross-checks that are done, and the precautions that are taken. Even so, some oversight some-where is always possible. On the other hand, it is difficult to see how two independent experiments could have made the same mistake. So perhaps the anomalies are real—and we are just missing a convincing explanation.

In this book we have argued that quantum mechanics is not an ultimate theory. Born's formula works only in the state of quantum death which we happen to be confined to. And quantum death came about sometime in the past by a process of quantum relaxation (Figure 58). If we go back far enough in time, we should reach a point when the universe was in its pre-quantum state—before quantum relaxation took place. At such early times, Born's formula will fail. According to inflationary theory the pattern of hot and cold spots in the CMB is a direct imprint of Born's formula in the early universe. So the implication is clear: a failure of Born's formula at early times can show up as anomalies in the CMB sky. Could the anomalies highlighted by the *Planck* team be telltale signs of what our universe looked like before quantum death?

Cosmic imprints of quantum relaxation

If the early universe was in a pre-quantum state, what might it have looked like? And how would it appear in the cosmic photograph known as the CMB? The early inflationary expansion is driven by a field called the 'inflaton field'. The quantum wrinkles in the inflaton field generate the primordial wrinkles that we see imprinted on the CMB. So the question is: was the early inflaton field in a pre-quantum state? And if so, did it carry a particular signature?

To develop some ideas we need to think about quantum relaxation for fields (instead of particles). This is easily done. It is convenient to split the field into its wavelength (or frequency) components and to think about one wavelength at a time (as we might do for a radio pulse). It turns out that a single wavelength—what physicists call a 'field mode'—behaves mathematically like a particle attached to a spring.[47] So quantum relaxation for the field is similar to quantum relaxation for a particle. An example is shown in Figure 64. In the top row we see the quantum equilibrium distribution—predicted by Born's formula—changing with time. In the bottom row we see a pre-quantum or nonequilibrium distribution moving quickly towards

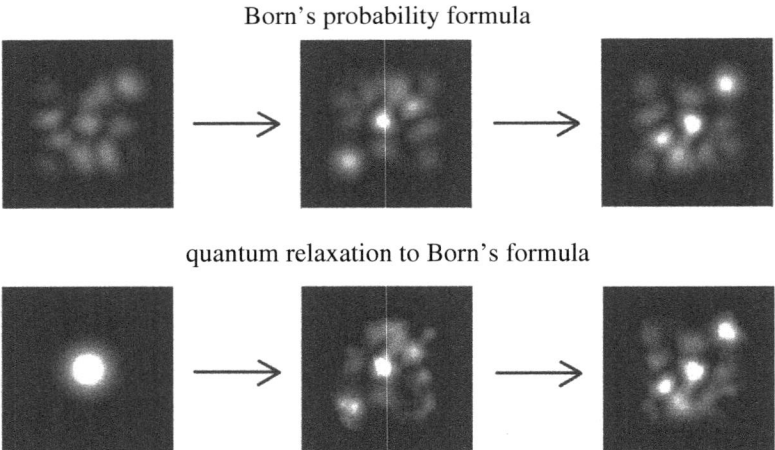

Born's probability formula

quantum relaxation to Born's formula

Figure 64. Quantum relaxation for a field on static (non-expanding) space.

---

[47] Technically, a particle in an oscillator potential.

equilibrium (or quantum death). The field rapidly relaxes to match Born's formula, just as particles do.

It might then seem that we have no chance of finding anything anomalous in the CMB, since fields in the early universe will quickly reach the equilibrium state described by Born's formula. But, actually, we have missed something out. For simplicity the computer simulation shown in Figure 64 does not take into account that space is expanding. The calculation would be valid for a field in the lab, where the expansion of space can be neglected. But in the early universe space is expanding very rapidly. And it turns out that this makes a difference.

An example is shown in Figure 65. The starting conditions are similar to those of Figure 64. But now space is expanding. And indeed it does make a difference. We see that quantum equilibrium is approached but not quite reached. The spread or width of the distribution remains smaller than that given by Born's formula (compare the final frames, top and bottom, on the right-hand side).[48] Quantum relaxation is slowed down or suppressed by the expansion of space. As a result the final state has *less noise* than a conventional quantum state—as measured by the reduced width of the probability distribution.

Born's probability formula

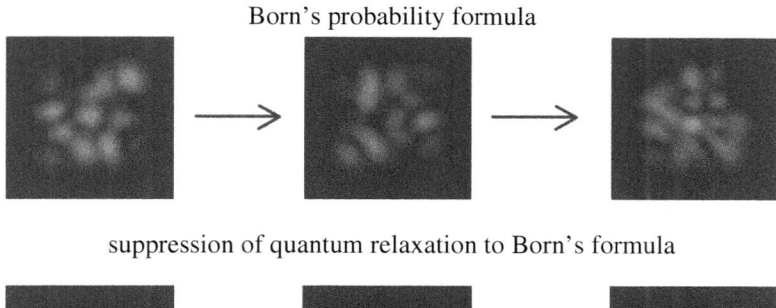

suppression of quantum relaxation to Born's formula

Figure 65. Suppression of quantum relaxation for a field on expanding space.

[48] These calculations were performed in 2013 in collaboration with Samuel Colin.

Quantum relaxation is suppressed on expanding space, but only at longer wavelengths.[49] At shorter wavelengths we find the usual efficient quantum relaxation. Recalling our analogy with coins (Chapter 5), it is as if the 'short-wavelength coins' are shaken thoroughly and reach an even ratio of heads and tails, while the 'long-wavelength coins' are shaken less thoroughly and do not reach an even mix.

What we have in Figure 65 is a computer simulation of quantum relaxation during a period of conventional (not exponentially rapid) spatial expansion. Something like this could have happened *before* inflation started, during what cosmologists call a 'pre-inflationary era'. This means that the final state shown on the right-hand side of Figure 65 is what the long-wavelength inflaton field may have looked like just before the beginning of inflation—with less noise than is usual in quantum mechanics. The long-wavelength components generate the power spectrum of hot and cold spots in the CMB at large scales. So the upshot is clear: we can expect to see a power deficit in the CMB at large scales—as in fact seems to be observed.

This conclusion is worth repeating. Computer simulations for the inflaton field on expanding space show that quantum relaxation is suppressed at long wavelengths. So we can expect to see Born's formula at short wavelengths and *violations* of Born's formula at long wavelengths. Specifically, there will be less noise at long wavelengths—which translates into less power in the hot and cold spots in the CMB at large scales. We seem to have a simple explanation for the large-scale power deficit in the CMB. The surprising (and still controversial) anomaly could be a telltale sign that the early universe was in a pre-quantum state.

And there is a bonus. If the long-wavelength field components break Born's formula, we can *also* expect them to break statistical isotropy—the noise or power will not be the same in every direction in space—again as seems to be observed at large scales in the CMB. For each field component the equilibrium state during inflation is symmetric (think of a perfectly round ball) and does not depend on direction. But a general nonequilibrium state breaks this symmetry (think of an egg shape) and can pick out a preferred direction. So we seem able to kill two birds with one stone: an early pre-quantum state can explain *both* anomalies at once.

At this point we might be excited. Perhaps the CMB data at large distances are showing us what the universe was like before it reached a

---

[49] Technically, at wavelengths larger than the 'Hubble radius'.

state of quantum death. But let us be careful and cautious—and sceptical. Perhaps the anomalies are caused by something else. Or perhaps the anomalies are not really there at all. To find out, we have to look more carefully at the data and compare with the detailed predictions of pilot-wave theory.

A failure of quantum mechanics?

Have we already seen a breakdown of quantum mechanics at large distances in the cosmic microwave background? To find out, we need a more precise idea of what to look for. The CMB data are hugely complicated. Any anomalies they may contain will not simply pop up in our faces. We have to know exactly what we are looking for, and then run demanding statistical tests (taking up huge amounts of high-speed computer time) to see if there is any evidence for that in the data. So what should we be looking for?

At long wavelengths we expect Born's formula to start breaking down. Specifically, the true probability distribution will have a reduced width, and there will be less noise than expected from quantum mechanics. But will this happen suddenly at a certain distance, or gradually as the distance increases? And if it happens gradually, can we say something more precise? Without a mathematical specification of what we are looking for, we will not be able to find it.

So here is what we will do. Let us rerun the computer simulation shown in Figure 65. But now let us try it at different wavelengths—some shorter, some longer. What do we find? At shorter wavelengths the results look more like Figure 64: the final distribution matches Born's formula, or nearly so. At longer wavelengths the results look more like Figure 65: the final distribution is 'squeezed' and the noise is reduced. For example, in some simulations the final noise is only about 70% of what it should be according to quantum mechanics (where noise is measured by the width of the distribution). And the longer the wavelength, the lower the noise. The results—found in collaboration with Samuel Colin—approximately fit the smooth curve shown in Figure 66.[50]

As mentioned, these simulations really tell us about the noise deficit just before inflation starts. To deduce the deficit during inflation, and

---

[50] Technically, this is an inverse-tangent function with three parameters. The actual results include small oscillations around this curve, which we can ignore in a first approximation.

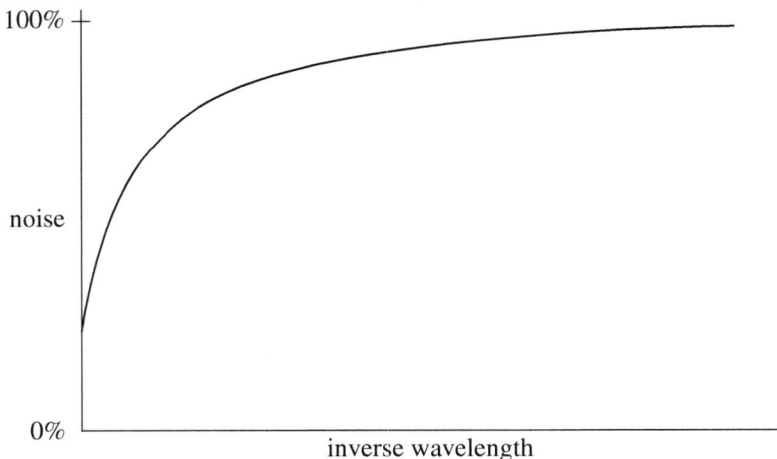

Figure 66. The noise goes down as the wavelength goes up. Here we plot the noise against the inverse wavelength (technically, the 'wavenumber'), so longer wavelengths appear on the left.

from there to find a prediction for the power deficit in the CMB, strictly speaking we need detailed modelling of the transition that marks the onset of inflationary expansion. The physics of that transition could well change the noise deficit. But such modelling is difficult and in any case has not yet been done. So, to get a preliminary prediction, we make a simplifying assumption: that the deficit will not change much during the transition. To what extent this turns out to be true remains to be seen.

So, with that simplifying assumption, we now have something definite to shoot for in the data. The *shape* of the CMB power deficit should look something like the curve shown in Figure 66. Can we find such a curve in the data?

An extensive search has been carried out in collaboration with French cosmologist Patrick Peter and Brazilian cosmologist Sandro Vitenti. We found that the reduced noise curve shown in Figure 66 fits the data fairly well. But the statistical significance of the fit is more or less the same as in quantum mechanics. In other words: the data neither support nor rule out this curve.[51] Because the data are so sparse at large scales, we are unable

---

[51] Technically, the significance of the fit takes into account the larger number of parameters. So the model could have been statistically disfavoured compared to quantum mechanics. Instead it performs comparably well—a modest success of sorts.

to see clearly the shape of any deficit that may exist there. It is difficult to draw clear conclusions. Even so, there are some intriguing hints.

The data show some preference for a curve like that in Figure 66 but with the reduced noise going all the way down to zero on the left-hand side. This corresponds to negligible noise at very long wavelengths. Now what we see at very long wavelengths corresponds essentially to what the universe was like before there was any quantum relaxation (since relaxation is almost completely suppressed at those wavelengths). Tantalisingly, this suggests that the universe may have started with no significant noise—essentially *all* the quantum noise we see today might have been generated by quantum relaxation. But, we emphasise, there is no more than a hint of this in the data.

At the same time, our search for a smooth curve like that of Figure 66 threw up a surprise: tentative evidence for a *sudden* (or sharp) drop in power beyond a certain wavelength. What the cause might be is unknown. And again, unfortunately, the data are too sparse for definite conclusions to be drawn.

At this point the reader might be getting impatient. Do these anomalies even exist? Is there significant evidence for a breakdown of quantum mechanics in the early universe—or is this a load of fluff? Well, if inflationary cosmology is true, it gives us an unparalleled opportunity to test quantum mechanics in the earliest moments of the Big Bang. But the most interesting place to look, at large distances, is exactly where the data are sparser and harder to interpret. That is just the way it is and we have to make the best of it. No one said this was going to be easy.

Readers who are old enough to remember the 1970s might be reminded of the grainy black-and-white photographs which appeared in British and American newspapers reporting on the search for the 'Loch Ness Monster'. Eyewitness accounts had reported spotting some sort of huge creature swimming in a lake or loch in a remote area of Scotland. It was even suggested it could be a living dinosaur. American polymathic inventor Robert Rines led the hunt for 'Nessie'. Taken with underwater cameras, the blurry images that appeared in the newspapers were tantalisingly unclear. Was that grey streak the long neck of a monstrous creature, or simply a piece of floating underwater debris, or only a meaningless shadow? No one could really say.

In the case at hand, we are peering through the mists of the cosmic microwave background as far back in time as we can possibly go, and at the very limits of what can be seen. Some of our best experimenters say they see anomalies and worry that something is wrong. Others say it is meaningless noise and we should sleep easily at night. Where lies the truth? At the time of writing, no one really knows.

The mystery of polarisation

With those caveats in mind, let us say a bit more about the results of our trawl through the CMB data. Something else came up, something as intriguing as it is difficult to evaluate. To describe it, we first need to explain more about the CMB.

We have talked about the temperature wrinkles of the background radiation. But as well as having a temperature (which varies slightly over the sky), the radiation also has a 'polarisation'—which means it vibrates in a preferred direction. Like light, or any electromagnetic wave, the radiation is made of vibrating electric and magnetic fields. If these fields vibrate in a preferred direction we say that the radiation is polarised. Now remember how the CMB was formed: by scattering off an early sea of charged particles. Basic physics shows that the scattered radiation should be polarised. Remembering also that the early universe is not perfectly smooth, we can expect the polarisation of the CMB to vary slightly across the sky. We should have polarisation wrinkles as well as temperature wrinkles. And indeed that is what we observe.

So there are really two main 'datasets' for the CMB: the temperature wrinkles *and* the polarisation wrinkles. Both datasets tell us something about the primordial perturbations—the small, mysterious deviations from a perfectly smooth early universe. Now we can guess where this is going. A full analysis of the CMB needs to look at both datasets. It is as if we have two different photographs of the early universe, and a complete picture needs both.

So now we can return to our trawl through the CMB data. We tried fitting the data to various anomalous reduced noise curves—the curve shown in Figure 66, as well as other examples for comparison. The purpose of these curves is to model the apparent power deficit at long wavelengths. As people often do when fitting CMB data, for simplicity we first looked at the temperature data only. We found an encouraging fit, with a modest degree of significance. It was certainly not enough to declare a discovery, but enough to warrant further study. Maybe we were on to something. We then added in the polarisation data, hoping for an even better fit. Instead the fit got worse. There are various kinds of polarisation data for the CMB, and the more polarisation data we added the worse the fit became. Adding polarisation data degrades the fit. The modest significance becomes negligible. It is as if some sort of shape is dimly discernible in the 'temperature photograph', but it disappears when we add in the 'polarisation photograph'. What is going on?

There is a simple and depressing answer. The shape we seem to see in the temperature photograph is not really there; it is merely a chance arrangement of meaningless noise. That would explain why it disappears when we add more data. As a simple analogy, imagine we toss a coin and get heads five times in a row. We might suspect the coin is biased or unfair. But if we toss the coin some more, maybe another twenty or thirty times, and we see more or less even numbers of heads and tails, we will conclude that those first five 'anomalous' results were just a fluke and we really have a fair coin. Similarly, for the CMB, maybe the temperature data alone happen to contain some fluke results of no real significance. So perhaps there is nothing to see here and we have wasted our time.

But this answer is not entirely convincing. We also tried adding other kinds of cosmological data (unrelated to CMB polarisation). If this answer was right, adding other kinds of data should also degrade the fit. But it did not happen. We found a degradation *only* when adding polarisation data. Treading cautiously, in a terrain where data are sparse and definite conclusions hard to come by, we might begin to wonder: if adding polarisation data systematically degrades the fit, could this be telling us something?

Here is an idea. Recall that the early universe is not perfectly smooth. There are primordial perturbations, which are imprinted on the CMB. These perturbations actually come in two types (called 'scalar' and 'tensor'). The first type is the main one, and describes small non-uniformities in the curvature of space. The second type is less important: it describes ripples in the curvature of space (called 'gravitational waves'). Now in inflationary cosmology the statistics for both types of perturbations are calculated with Born's formula. Because they have a common origin, the two kinds of statistics turn out to be related. Cosmologists call this a 'consistency relation'. This relation must be true if inflation is true. Or at least: that is what quantum theory tells us. But if we allow Born's formula to be broken, there is *no* reason why the statistics for the two types of perturbations should be related. Going back to our analogy with coins, imagine that we have two different types of coin—say copper and silver. If both types show an uneven ratio of heads and tails, there is no reason why they should both show the *same* uneven ratio. For example, the copper coins might show more heads than tails while the silver coins might show more tails than heads. In the same way, if Born's formula is broken, the nonequilibrium statistics for the two types of perturbations (scalar and tensor) will generally be different. This means that the usual consistency relation is only a feature of quantum death. Outside of quantum death, that relation will be broken.

What does this have to do with the polarisation data? The point is that the second type of perturbation is expected to have a small effect on the temperature wrinkles but could have a somewhat larger effect on the polarisation wrinkles. In that case, when we add in the polarisation data, we are effectively adding in the second type of perturbation. And so now it is interesting. If the two types of perturbation break Born's formula in different ways, adding them together might confuse the signal. Think again of our copper and silver coins. If the copper coins show more heads than tails, while the silver coins have it the other way around, if we simply lump them all together then overall we could find a deceptively even mix.

We seem to have the beginnings of an explanation for why the fit becomes degraded when we add in polarisation data. There are two independent violations of Born's formula in the early universe, one for each type of perturbation, with two different reduced noise curves in play.[52] When we try to fit all of the data to only one curve, the results are confused and the signal seems to disappear. At the time of writing this is only an idea. To find clear evidence for this we would need to rerun all the fits with two different reduced noise curves. This could be done but has not been tried yet. What the results might be is anyone's guess.

To conclude, it seems fair to say that pilot-wave theory predicts the right *kinds* of anomaly in the CMB at large distances: a noise deficit, and

---

[52] Technically, two different sets of parameters defining the class of curves shown in Figure 66.

wrinkles whose statistics depend on direction. It also predicts 'inconsistent' primordial perturbations. And it provides one unified account for all of the above. This is encouraging. But to prove or disprove the theory we need more precise predictions and more statistical tests. It is difficult work. And there is still a long way to go. Are there genuine anomalies in the early universe? Possibly. Have we found a breakdown of quantum mechanics at the beginning of time? No one knows.

If it sounds as though we are struggling, that is because we are. And we are not alone. In 2020 the *Planck* team published a new analysis of the polarisation data. It seems to show little or no sign of any anomalies—in apparent contradiction with the temperature data. Two different snapshots of the early universe do not quite match up. Perhaps a future space mission, specially designed to study the CMB polarisation, will resolve the mystery. But that may be ten years away. For now, we leave the last word to the *Planck* team:

> . . . nature might be trickier than we imagine. As yet, there is no convincing hypothesis for what kind of new physics could be causing the anomalies. So, it could be that the phenomenon responsible only affects the temperature of the CMB, but not the polarisation. . . . In the meantime, the mystery of the anomalies continues.

7

Relics from the Early Universe

Quantum archaeology

In the last chapter we saw that the early pre-quantum universe could have left traces in the cosmic microwave background (or CMB). But whether there really are such traces is difficult to prove. There are reports of anomalies, of the general form expected from pilot-wave theory, but the data are currently too uncertain to draw firm conclusions. It is all so tantalising—and frustrating. It may be another ten or fifteen years before future space missions can finally settle these questions.

In this chapter we explore a different route out of quantum death. Instead of looking for evidence of past violations of quantum mechanics, we will be hunting for particles that *today* violate quantum mechanics. These particles could be relics from the early universe, which failed to reach the state of quantum death, and which have so far evaded detection or not been noticed. As we will see, space could be filled with particles in a pre-quantum state. They might be streaming through the air around us right now. If only we can figure out how to spot them. We saw in Chapter 5 that such particles could be deployed to work what most physicists today would regard as miracles: beating the uncertainty principle and sending superluminal signals. As we will show in Chapter 10, the discovery of such particles would herald a technological revolution whose outlines we can only dimly foresee. If subquantum matter were discovered, it would surely be the most precious material known to humanity. In a word, we are looking for a buried treasure of almost miraculous powers.

We should say right away that finding particles left over from the early universe is the easy part. After all, everything we see around us is left over from the early universe. But what we really want are particles which have been essentially untouched since early times. That, too, is not as difficult as it sounds. The photons (or energy packets of light) making up the

Beyond the Quantum. Antony Valentini, Oxford University Press. © Antony Valentini (2025).
DOI: 10.1093/oso/9780198853749.003.0008

CMB are an example. They are 'relic cosmological particles' from early in the history of the universe and which have been streaming essentially freely through space ever since. And there is every reason to think that the universe contains other—more exotic—relic particles from even earlier times.

But why bother with more exotic relics when we already have CMB photons? To see why CMB photons are of no use to us, remember from Chapter 6 that when the early sea of charged particles combined to form neutral atoms of hydrogen, the universe became transparent to radiation and from then on the CMB photons could stream freely. The photons 'decoupled' from matter, meaning they no longer interacted significantly, which is why they remain with us today as a record of those early times. Unfortunately, the CMB photons did not decouple until around 380,000 years after the Big Bang, by which time they will have long since relaxed to the state of quantum death.

To find relic particles that violate quantum mechanics and Born's formula, we need particles that were created much earlier, at times closer to the beginning, and that have been streaming essentially freely through space ever since. Such 'primordial relics' are more likely to have evaded quantum death. According to our leading theories of cosmology and particle physics, there are reasonable prospects for finding such particles. Whether or not they really violate the laws of quantum mechanics is, of course, another matter.

So now we face two questions. Which of these primordial relics are more likely to have avoided quantum death? And how can we detect them today? As so often in the history of science, what we are looking for might be staring us in the face.

The mystery of dark matter

The past thirty years have seen the emergence of precision cosmology. The accurate maps of the tiny temperature wrinkles in the CMB, obtained by the *WMAP* and *Planck* satellites, have revealed the history of our universe in exquisite detail. Our knowledge of the universe now seems extraordinarily complete. Except for one thing. Bizarre as it may sound, we do not know what most of the universe is actually made of.

According to our best measurements and calculations, only about 5% of our universe is made of the ordinary matter we find in the lab. What about the rest? Well, about 68% seems to be made of something called 'dark energy', which is responsible for the observed acceleration of our expanding universe (the expansion is speeding up). Opinions vary over how mysterious this is. Some argue that it is just a 'cosmological constant'. This is a simple extra ingredient introduced by Einstein in 1917 to supplement his theory of gravity. It amounts to a sort of residual energy and pressure that is present even in empty space. Others try to explain dark energy in terms of a new physical field or other exotic effect from high-energy physics. As yet there is no consensus on what dark energy really is, and this remains an active area of research.

What about the remaining 27%? This is believed to be a form of matter which we cannot see. Cosmologists call it 'dark matter'. Remember that the matter we *can* see is visible to us only because it emits light (or other forms of electromagnetic radiation) or because it reflects light. We can see the Sun and stars because they are nuclear-powered fires, and we can see the planets and the Moon because they reflect light from the Sun. We might call the matter we can see 'bright matter'—though no one uses that term. Now it seems there is also a mysterious dark matter out there in space, which does not emit light and does not reflect it either. And there are many theories about what exactly it might be. However, before getting into those theories, let us first look at why cosmologists believe dark matter exists in the first place.

There are literally billions of galaxies populating our universe. And each galaxy is a conglomeration of billions of stars. In fact galaxies come in a variety of shapes and sizes. An example of a 'barred spiral' galaxy is shown in Figure 67. But what holds a galaxy together? Why do the stars not fly off freely into intergalactic space? The answer, of course, is gravity. The earth and other planets in our solar system do not fly off into space, but remain in orbit around the Sun, because of the gravitational force from the Sun. Similarly the stars populating a galaxy remain more or less confined to where they are because of the overall gravitational pull which they exert on each other. Like the planets in the Solar System the stars move around on various orbits, but overall they remain confined to the galaxy they are in. In the jargon, a galaxy is a 'gravitationally bound system'.

Or at least, that is the theory. But when we examine the details, it all falls apart—quite literally. Some of the stars are moving so fast they should be able to escape the galaxy's gravitational pull—just as a theoretical bullet shot fast enough into the sky (at a speed exceeding eleven kilometres

Figure 67.   The barred spiral galaxy NGC 1300 as seen by the Hubble Space Telescope.

per second) can escape the earth and fly off into space. We can work out the total mass of a galaxy by counting up the number of stars it contains. We can then apply the laws of gravity to calculate the gravitational force exerted by the galaxy as a whole. That force should be enough to hold the galaxy together and to prevent the most rapidly moving stars from escaping. But it does not work out that way. For most galaxies the total mass is not enough to hold it all together. It is as if there must be some invisible extra 'glue' keeping the whole thing in one piece. What is going on?

There is more. Many galaxies, including our own, are rotating. In the outer regions there are stars and gas in orbit around the centre. We can apply the laws of gravity to work out how fast they should be moving. This problem was solved long ago by Kepler and Newton for planets in our solar system. Because gravity becomes weaker further away from the Sun, distant planets should orbit more slowly. And indeed they do. Similarly, stars and gas further away from the centre of a galaxy should orbit more slowly. But they do not. Instead, further out the orbital speed remains more or less the same, seemingly in complete violation of the most basic laws of physics. Again, what is going on?

There is a simple answer to both problems: dark matter. When we calculate the mass of a galaxy by counting up the visible stars, we are making a mistake. There is in fact more matter there, only we cannot see it. And that extra matter provides an extra gravitational force, which is enough to stop those rapidly moving stars from escaping. Not only does the galaxy we see contain dark matter within it, there is also an invisible 'halo' of dark matter extending far beyond the visible part of the galaxy—and this explains why the orbital speeds do not become smaller as we go further out. For all this to work, however, we need not just a small amount of dark matter, but lots of it. In fact, for an average galaxy, there has to be about five times more dark matter than visible matter.

From the basic physics of galaxies, then, the evidence for dark matter is overwhelming. With one caveat: we are assuming the usual laws of gravity as given by Newton and Einstein. According to those laws, there is not enough mass in a galaxy to hold the stars together. But what if those laws are wrong? For example, perhaps the gravitational force is stronger than we think, not because there is invisible dark matter, but because the laws of gravity are different from those written in the textbooks. According to some alternative theories, there is no dark matter and instead our laws of gravity are wrong. This is a reasonable possibility and might turn out to be true. But at the time of writing the most likely explanation is still widely believed to be dark matter.

Further evidence for dark matter comes from cosmology. As we saw in Chapter 6 the statistics of hot and cold spots in the CMB contain a huge amount of information about the past history of our universe, including details of how it expanded over time. The expansion of the universe is affected by the matter and radiation it contains (according to Einstein's theory of gravity). And so by measuring the CMB we can work out the past expansion and deduce what the universe contains. And the results point to a seemingly clear conclusion: 5% normal matter, 27% dark matter, and 68% dark energy.

So let us accept, as most cosmologists do, that our universe contains a huge amount of unseen dark matter. What might it be? There is a veritable zoo of competing theories, some more exotic and speculative than others. It has been suggested, for example, that dark matter could be made of 'brown dwarfs', which are a sort of failed star that is not massive enough to ignite. Alternatively, it could be made of black holes formed from collapsing stars. But most astronomers believe that such 'massive compact halo objects' (known in the trade as MACHOs) are unlikely to be sufficiently abundant to account fully for dark matter. They certainly exist but there are not enough of them. So let us look instead at another idea, which is widely regarded as more likely: that dark matter consists of relics from the early universe. We can call it 'relic dark matter'. What might it be made of?

One possibility is that relic dark matter is made of 'primordial black holes' which formed in the early universe. These could be much smaller and much more abundant than the black holes created later by collapsing stars (Chapter 9). Remember that the early universe is very hot and very dense—and not completely smooth. It has wrinkles, which are usually quite small. But here and there the wrinkles will be larger than average. And in some places they might be so large that their own gravity makes them collapse to form a black hole. As first theorised by Hawking in 1971, this is likely to have happened to some extent in the early universe. The question is: how many primordial black holes have survived to the present day and with what masses? And are they enough to account for dark matter? This is an active area of research with as yet no final consensus. It is a reasonable idea and might turn out to be true. But we do not yet know.

Here is another reasonable idea. Perhaps the hot Big Bang created exotic high-energy particles that interact so weakly with ordinary matter that today we are unable to see or detect them. Those particles are still floating around in space. They can pass through the earth without touching anything. But they are massive and subject to the force of gravity. This means they can form a halo around galaxies—like dark matter. And so dark matter might be made of relic particles from the early universe. Because they are weakly interacting and massive they are known, appropriately enough, as 'weakly interacting massive particles' or WIMPs. Speculative as they may sound, and despite doubts raised in recent years by some theorists, WIMPs remain the leading candidate for dark matter.

Survivors of quantum death?

We seem to have strayed a long way from quantum mechanics. What has dark matter got to do with escaping quantum death? The answer is that it might conceivably be the buried treasure we are looking for. Recall that CMB photons are of no use to us because they decoupled (stopped interacting with other things) far too late: they will have already reached quantum death. We need something that was created much earlier, and that decoupled much earlier. Depending on the details, dark matter might turn out to be just what we need.

To understand how this can work, let us recall Chapter 6 where we looked at inflationary cosmology (the leading theory of the early universe). We saw that, at very early times, the universe had a brief period of exponential expansion driven by an inflaton field, whose quantum wrinkles are now imprinted on the CMB. One thing we did not talk about is how the early inflationary expansion comes to an end. This happens when the inflaton field loses its 'potential energy' and is no longer able to drive the expansion. Think of the field as a ball rolling down a hill. Near the top of the hill the energy of the ball drives the rapid cosmic expansion, but once that energy is lost the expansion slows down. Then something crucial happens. When the ball reaches the bottom of the hill, it bobs up and down (or oscillates) before coming to rest, and the energy of the inflaton field is converted into matter and radiation (in a process whose fine details are perhaps not fully understood). The result is a huge rise in temperature, which in effect causes what we usually think of as the hot Big Bang. The matter in our universe was created by this early process—called 'inflaton decay'.

So let us linger on that thought for a moment. The matter in our universe was created at early times by the decay of the inflaton field. But we also saw in Chapter 6 that the inflaton field can carry primordial quantum nonequilibrium, or primordial deviations from Born's quantum formula (which might show up as anomalies in the CMB). Putting those two thoughts together, we realise that in the early universe the decaying inflaton field can create nonequilibrium particles which violate Born's formula. In fact, the early hot universe could be filled with such pre-quantum particles.[53]

What happens then? If the particles are interacting they will quickly relax to quantum death. Any deviations from Born's formula will disappear long before the particles complete the long journey from the Big Bang to us. The radical new physics we are looking for will be erased forever. But, if some of the particles are *not* interacting, they could avoid quantum death and continue to violate the laws of quantum physics. And so we catch a glimpse of how our sought-for buried treasure could conceivably survive untouched to the present day.

What we need is quite clear: particles that were created at early times *and* that have hardly interacted with anything else since the time of their

---

[53] This scenario has been studied in collaboration with the British physicist Nicolas Underwood.

creation. This may sound like wishful thinking. In fact, according to some prominent theories, dark matter is made of precisely such particles.

Here is an example. A theory of particle physics known as 'supersymmetry' predicts that new kinds of particles will be created at sufficiently high energies, particles that are in a sense partners of the particles we are familiar with at lower energies. These new 'superpartner' particles ought to be copiously produced in the early universe. The precise predictions depend on the details of the model, concerning which there are many unknowns. One popular scenario has been much studied by theorists in recent decades. In the early universe, at very high temperatures, a superpartner called the 'gravitino' is created in large numbers. These particles interact so weakly that from the time of their creation they hardly interact with anything else (if certain conditions are met[54]). On some estimates, a large fraction of the present-day mass of the universe could take the form of relic gravitinos. They could be passing through the air around us right now without us noticing. The only force they feel significantly is gravity. This means they can still cluster into a halo around galaxies. In other words, according to these scenarios, dark matter could be made of relic gravitinos—or of some other, comparable particle predicted by supersymmetry.[55]

Literally hundreds of scientific papers, and not a few PhD theses, have been written about theories of gravitinos as dark matter, or about some other superpartner particle as dark matter. Details vary from one scenario to another, depending on the precise details of particle physics at high energies. But the general gist is clear. Dark matter could be made of exotic particles that were created in the early universe and which have hardly interacted with anything else since. This is just what we need if we are to have a realistic hope of finding particles today that continue to violate quantum mechanics.

At this point the reader might object. If the relic dark matter particles do not interact with anything else, they will pass through our equipment unnoticed. So how can we possibly detect them or do anything useful with them? We seem to be in a terrible fix. On the one hand we need particles that do not interact with anything else, to have a hope of them avoiding quantum

[54] Technically, they must be created at a time when the ambient temperature is lower than their decoupling temperature.

[55] In recent years supersymmetry has become somewhat disfavoured by particle physics experiments. Still, it remains a viable and popular theory, and in any case we are only citing it as an example.

relaxation and quantum death. On the other hand we need particles that interact with our equipment, to have a hope of finding them. We cannot have it both ways. So it seems that our quest for pre-quantum particles was doomed from the start. Unless . . .

Disintegrating dark matter

. . . unless the relic particles are unstable and eventually disintegrate into something else that we *can* detect. Again this may sound like wishful thinking. But it is a straightforward prediction of many prominent theories of dark matter. In some models the relic particles can 'decay' to produce X-rays. In other models the relic particles can collide and 'annihilate' to produce gamma rays (short-wavelength X-rays). A pair of gravitinos, for example, is expected to annihilate into a pair of gamma-ray photons. The point is that we do not need to detect the dark-matter particles themselves. Instead we can try to detect the X-rays or gamma rays that are produced when those particles disintegrate. If the disintegrating particles violate Born's formula, so will the products.

At this point we need to talk a bit about how pilot-wave theory accounts for physics at high energies, where particles can be created and destroyed (as routinely observed in particle accelerators). We touched on this topic in Chapter 4, when we discussed the pilot-wave theory of the electromagnetic field. There we saw that photons are not really particles but rather energetic excitations of the electromagnetic field. In pilot-wave theory there is a definite field that evolves with time. As the field changes, its energetic excitations can rise and fall, amounting to the creation and destruction of particle-like photons. This occurs, for example, when an atom emits or absorbs radiation (as was originally discussed by Bohm). Now, at high energies, other kinds of particles can also be created and destroyed, and we must ask how pilot-wave theory can describe this.

There are in fact two alternative approaches. In the first, *all* particles are regarded as energetic excitations of fields (as for photons). The creation or destruction of particles corresponds to different frequencies of the field being excited or de-excited, much as in the case of the electromagnetic field. In the second approach, particles such as electrons[56] really are particles moving around in space, as described earlier in this book for physics at low energies. However, even so-called empty space contains

---

[56] Technically, all 'fermions'.

large numbers of them—filling a 'negative-energy sea', along lines orig-inally proposed by Dirac in 1930. If one of the negative-energy particles is excited to a positive-energy state, we see it as an electron, while the 'hole' left behind in space appears to be a positron (the oppositely charged 'antiparticle' to the electron). We can then explain what appears to be the creation of an electron–positron pair out of the vacuum. By the same token, an electron–positron pair can annihilate into pure radiation. Similar reason-ing applies to other kinds of charged particles. In this way, Dirac was the first to predict the existence of what we now call antimatter. Pilot-wave theory gives us a more precise version of Dirac's original idea.

Which of these pictures is correct—with fields only, or with a Dirac sea of particles—will probably not be known until we discover quantum nonequilibrium, which will enable us to find out by probing the details of the underlying reality. In either case, pilot-wave theory is able to describe the creation and destruction of particles, and in particular it can describe the possible disintegration of dark matter. In both pictures, if the initial state (of fields or particles) violates Born's formula, so will the final products.

All this means that, if we are lucky enough to detect X-rays or gamma rays produced by disintegrating dark matter, our quest for a breakdown of quantum mechanics may be at an end. We will have in our hands something that originates, albeit indirectly, from the very beginning of the universe, and which has not been tampered with since. We would then need to test the X-rays or gamma rays to see if they obey Born's quantum formula.

That may sound all well and good, but how might we go about finding X-rays or gamma rays produced by disintegrating dark matter? In fact, an ongoing high-profile experiment may have already found what we are look-ing for. The *Fermi Gamma-ray Space Telescope*, launched in 2008 and still operating, has detected a puzzling 'excess' of gamma rays coming from the centre of our galaxy (Figure 68). This has been confirmed by data gath-ered over several years. The jury is still out on the source of the excess, but one of the leading contenders is disintegrating dark matter. Several studies support the idea and aim to narrow down the mass of the disinte-grating particle—recent estimates put the mass at about 50 or 60 times the mass of a proton. But some astronomers remain sceptical and favour more conventional astrophysical explanations. In any case, if disintegrating dark matter does exist, there is a fair chance that the *Fermi* telescope—or other comparable experiments—will find it.

So there we have it. According to our reasoning there is a case for testing the Galactic excess gamma rays for violations of the usual laws of quantum

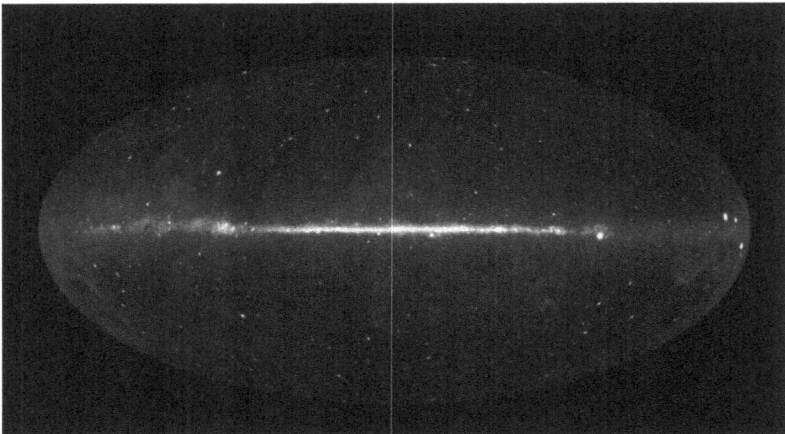

Figure 68.   An unexpected surplus of gamma rays from the centre of our galaxy (brighter areas) as seen by the *Fermi Gamma-ray Space Telescope*. The rays are widely thought to be created by disintegrating dark-matter particles left over from the early universe. Might these rays violate the laws of quantum mechanics?

mechanics—and of Born's formula in particular. There are various ways of doing this. We could allow the incoming rays to hit an appropriate two- or multi-slit barrier (perhaps a diffracting crystal or grating) and form an interference pattern. If the rays violate Born's formula the usual wavy pattern will be blurred or modified in some way. Alternatively the rays could be allowed to scatter off electrons or atomic nuclei. If Born's formula is broken we will see anomalies in the proportion of rays that come out at different angles. We could also look for anomalies in polarisation statistics (the proportions of rays vibrating in different directions). In fact any of the usual quantum predictions made with Born's probability formula can break down.

Testing Born's formula in space

Such experiments could not be carried out in the lab though. Incoming gamma rays from space are strongly absorbed by the earth's atmosphere, which is why gamma-ray telescopes such as *Fermi* are housed on orbiting satellites and not on the earth's surface. A future space mission would have to include quantum experiments to test the incoming gamma rays for violations of Born's formula. There is no obstacle of principle to doing this. Quantum experiments in space are already being planned to test quantum

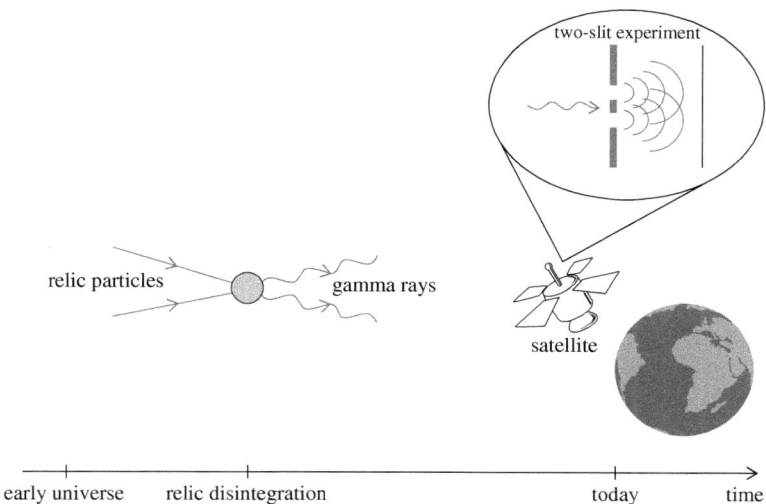

Figure 69. A breakdown of quantum mechanics? According to prominent theories of particle physics, an invisible sea of relic particles pervades our galaxy and has remained virtually untouched since the earliest times. Eventually the particles disintegrate into gamma rays, which could violate Born's formula—producing anomalies in a two-slit experiment.

entanglement over large distances (Chapter 10). Here we are proposing a quantum experiment aimed at testing Born's formula, and specifically for photons making up the observed gamma-ray excess from the centre of our galaxy. A dedicated satellite could, for example, include an onboard two-slit experiment to test Born's formula for incoming photons (Figure 69).[57]

Similar experiments might be done with other suspicious X-ray or gamma-ray sources from space. There is, however, a catch to look out for. If the incoming X-rays or gamma rays violate Born's formula, then our equipment might not be able to see the rays properly.[58] Astronomers are searching for a 'sharp spectral line'—photons with a specific energy or frequency—produced by disintegrating dark matter. If the photons violate Born's formula, the detected line could look strangely distorted, with

---

[57] The QUICK³ satellite mission includes a more elaborate interferometer designed to test Born's formula in space.

[58] This has also been studied in collaboration with Nicolas Underwood.

a seemingly impossible shape. The shape of the line might also depend on which type of detector is being used, which would confuse things even more. Some odd anomalies seemingly of this kind have been reported, but we are a long way from being able to draw any conclusions. The reported anomalies are likely to have other causes. Still, in our search for violations of quantum mechanics in radiation from disintegrating dark matter, we need to bear in mind that the search for disintegrating dark matter might itself be confused by such violations. Again, no one said this was going to be easy.

Despite such uncertainties, however, what matters in the end is this. If we observe radiation from space that fails to produce the usual wavy pattern in a two-slit experiment (Figure 69), then clearly quantum mechanics will have failed. If our ideas are right, a new window onto subquantum physics will have been opened. In the meantime, the search for disintegrating dark matter provides us with some clues for where to look. At the time of writing, the excess gamma radiation from the centre of our galaxy seems like a good place to start.

All this turns, of course, on the idea that dark matter is in fact made of relic particles from the early universe (as thought by many theorists). But that idea could turn out to be mistaken. Perhaps dark matter is not made of particles at all. In that case searching for dark-matter disintegration seems futile. Though not quite. As we have mentioned, another popular theory holds that dark matter is made of primordial black holes—large numbers of tiny black holes created in the early universe. As first understood by Hawking in 1974, because of quantum effects black holes can 'evaporate' and eventually explode in a burst of radiation that includes gamma rays. In other words, primordial black holes can also disintegrate to produce something we can see. Satellites such as *Fermi* might detect such events. But is there any reason why radiation emitted by black holes could violate the laws of quantum mechanics? There is indeed. But to explain it we first need to look at the fraught relationship between quantum mechanics and Einstein's theory of gravity.

8
Saved by Gravity

Trapped forever in a quantum fog?

The central thesis of this book is that we are trapped in a state of quantum death. Everything we can see and touch is imbued with a statistical noise, a sort of quantum fog that makes it impossible for us to see and control the finer details of microscopic reality. Your friend who studied physics at university was told that this is because of an unbreakable law of physics called the uncertainty principle, which sets fundamental limits on what we can see and measure. We have argued that this is not correct. It does not have to be that way. Quantum uncertainty is not a law of physics but merely a peculiarity of quantum death. Our world is pervaded by quantum fog not because of any fundamental law or principle but because the fog was created by the violence of the early universe. If we go back far enough in time we will find a pre-quantum era, before the fog was formed, a state of quantum life in which reality is clearly visible and there is no uncertainty. At such early times, the fog clears and Born's probability formula fails.

We worked hard at this in Chapters 6 and 7, peering back in time as far as humanly possible, looking for a failure of quantum mechanics. We found some tantalising hints, but as yet there is no hard evidence. The truth is, even if the ideas described in this book are correct, we will need to be lucky to find evidence for them. For all we know the pre-quantum era could have been so comprehensively buried in the violence of the early universe as to have vanished without trace forever—leaving us trapped in a quantum fog with no means of escape.

For instance, in Chapter 6, we saw how early quantum noise is imprinted today in the cosmic microwave background (or CMB), giving us hope that an early failure of Born's formula could be revealed as anomalies in the CMB. And anomalies of the right kind have been reported. But those anomalies might turn out to be just meaningless noise in scanty data. Or they might be real, but have some other cause. Here is a depressing thought:

Beyond the Quantum. Antony Valentini, Oxford University Press. © Antony Valentini (2025).
DOI: 10.1093/oso/9780198853749.003.0009

perhaps an early failure of Born's formula is imprinted on the CMB, but only at distances so large as to be out of our reach. That is entirely possible if the very early (inflationary) universe expanded so much as to push the pre-quantum anomalies out to extremely long wavelengths. In other words: even if our theory is correct, we might never be able to prove it, because the anomalies might be at distances too large for us ever to measure in practice.

Much the same is true for our attempts, in Chapter 7, to find relic particles from the Big Bang that violate Born's formula. Perhaps such particles did survive for a while, but eventually succumbed to quantum death (for example, because of small interactions with other particles). Or perhaps they still exist today somewhere in the vast universe but are too far away for us to ever find them.

For all sorts of reasons, then, our ideas might be correct but unprovable. This is the stuff of scientific nightmares—unless we are missing something.

The reader may well have wondered: surely there must be an easier way. Why do we have to go picking our way through cosmic debris from the beginning of time, looking for nonequilibrium particles that violate Born's formula? Why can we not *create* such particles now in the lab, perhaps by hitting atomic particles very hard (Figure 70)? Or by some other, clever trick?

Figure 70. Knocking it out of equilibrium. Can we *create* quantum nonequilibrium by hitting atomic particles very hard? According to our current understanding, the unfortunate answer is no. It seems that we are trapped forever in a quantum fog. Unless we are missing something.

If only that were possible. We saw in Chapter 1 that our present state of quantum death is analogous to the futuristic nightmare known as the heat death of the universe, which some theorists worried about in the nineteenth century. Basic principles of thermodynamics show that, billions of years hence, everything in the universe will eventually reach thermal equilibrium—with the same temperature everywhere—and once that happens there is *no* way out. Similarly, the laws of pilot-wave theory show that, once everything in the universe reaches quantum equilibrium—with the same quantum noise everywhere—there is no way out.[59] That is the state of quantum death. Try as we might to reduce the quantum noise in some atomic system in the lab, we will fail because our equipment is imbued with the same noise. There is no escape from quantum fog. It is everywhere, in us, and all around us. Our only hope is to find something left over from the early universe, before the fog took over. If we cannot find it, we are forever doomed.

Unless we really are missing something.

The mystery of gravity and quantum mechanics

Let us talk a bit about gravity. We saw in the Prologue how Kepler and Newton understood that the planets are kept in orbit by a gravitational force that emanates from the Sun. Some two and a half centuries later this understanding was deepened, and radically changed, by Einstein. According to him, what is really happening is this. A massive body like the Sun warps the space around it. The space is no longer flat like the surface of a table, as taught in school by Euclid, but 'curved' like the surface of a sphere. When a body like the earth moves through that space, it takes the shortest path. But because the space is curved, the shortest path appears to be curved. And the result is a planetary orbit.

Einstein's theory of gravity is without doubt the most beautiful theory in the history of theoretical physics. And we can hardly do it justice in one glib paragraph. In fact what we have said is not quite accurate. It is not just space that is curved but 'spacetime'—a four-dimensional amalgamation of

---

[59] Actually, the 'fourth loophole' mentioned at the end of Chapter 1 may offer a way around this. Corrections to de Broglie's law of motion, associated with nodes of the quantum wave, could create nonequilibrium in certain violent or fast processes (such as high-energy particle collisions). But these ideas are not discussed in this book as they are still being developed.

space and time. And the shortest path is taken not in space but in spacetime. However, this distinction will not matter for our purposes.

Gravity is not like other forces. In fact, according to Einstein, it is not really a force at all. In classical physics, a particle left to itself moves freely and uniformly in a straight line (as required by Newton's first law). When acted on by a force, a particle *deviates* from uniform motion in a straight line. For example, an electric or magnetic force pushes a charged particle and makes it accelerate, so the particle has a changing speed and generally does not move in a straight line. Newton thought that something similar was happening to the Moon in its orbit around the earth, and to the earth in its orbit around the Sun. According to him, gravitational forces made these objects deviate from their natural motion in a straight line. But Einstein showed this was wrong: those objects *are* following their natural motion in a straight line, but in a curved space (or spacetime). In Einstein's theory, space is a physical object which can bend and stretch in myriad ways, like the surface of an elastic balloon. Objects moving through it wind their way along and around those distortions. And that is Einstein's explanation for gravitation (Figure 71).

So gravity is not merely another force like electromagnetism but something quite different—something radically and beautifully different. This raises a question: can gravity be described by quantum mechanics?

Figure 71. Einstein's explanation for gravitation. The Sun curves the space (or spacetime) around it. The earth moves in a straight line, following the shortest path, resulting in a planetary orbit around the Sun.

We should say first of all that conventional forces *can* be described by quantum mechanics. Though when we say 'described' we mean in the usual vague sense. For example, we know that an atom is somehow held together by electrical forces. Positively and negatively charged particles attract each other. In quantum mechanics there are no clearly defined particle trajectories, but we can define a quantum wave, whose squared magnitude tells us the probability for finding the charged particles here or there (Born's formula). This can be done for charged particles interacting by electromagnetic forces. The basic equations of this 'quantum electrodynamics' were written down by Dirac in 1927. It took about twenty years to sort out some technical difficulties, and by 1949 we were able to apply the equations to make practical calculations (thanks to work by Richard Feynman and Julian Schwinger in the USA, and Sin-Itiro Tomonaga in Japan, who in 1965 shared a Nobel Prize). The theory was valid for charged particles interacting at high energies, where particles can be created and destroyed.[60] Among other applications, it was possible to use Born's formula to calculate the probability for particles to be scattered in various directions after a collision. This success set the stage for what we now call 'high-energy physics'. We only had to find the analogous theory for the other forces, in particular the weak and strong nuclear forces. By the 1970s this had been accomplished, giving us what we still call the 'standard model' of particle physics. This model includes the famous Higgs boson. Its statistical predictions—calculated from Born's formula— have been abundantly confirmed by experiments at high-energy particle accelerators.

What about gravity? We have said that gravity is not like other forces. Even so, we might try to describe it with quantum mechanics. Here is how it should go. As usual we need to define a quantum wave. And, following Born's formula, the (squared) magnitude of the wave will tell us the probability for this or that outcome. The only difference is this. Instead of telling us the probability for particles to be here or there, the quantum wave will tell us the probability for the curvature of space to be this or that. In other words: instead of a vague statistical account of particles, 'quantum gravity' should give us a vague statistical account of curved space.[61] That

---

[60] How pilot-wave theory accounts for such processes was outlined in Chapter 7.

[61] Note that here, in what is technically known as 'canonical quantum gravity', the object to be described really is curved space and not curved spacetime.

would not be an account of quantum reality, of course, but it would be an important advance. It would accomplish for gravity what Dirac and others had accomplished for electromagnetism.

Throughout this book we have been critical of the quantum theory of particles and fields, and of atoms and molecules, that is taught to students in our universities and in our textbooks. Remarkably, when it comes to the quantum theory of gravity, the situation is even more disturbing.

Quantum gravity and the disappearance of time

Quantum gravity has a long and tortuous history. After a few decades of confusion, by the 1960s the American physicists John Wheeler and Bryce DeWitt had succeeded in writing down what appeared to be the correct equation for the quantum wave. It is known, appropriately enough, as the Wheeler–DeWitt equation. It is widely regarded as the fundamental equation of quantum gravity. It has a certain seductive simplicity. It seems to be inevitable. If we apply the usual rules of quantum mechanics, to Einstein's beautiful theory of gravity, this is the result. The mathematical logic is compelling. Whichever way we look at it, we end up with the Wheeler–DeWitt equation. So it must be correct?

But there is a problem. The equation makes no reference to time. Taken at face value, the equation seems to say that the quantum wave always stays the same—it is static, frozen, unchanging. This is in stark contrast with the quantum wave as usually understood. For example, for a single particle the quantum wave sloshes around in space rather like a water wave on the surface of a pond. It certainly does not stay the same. It moves and changes (for example, when it hits a two-slit barrier, Figure 27). Whereas, if we include gravity in our account of a quantum system, it seems that the quantum wave does *not* change.

Let us pause and think through the implications. Remember that the squared magnitude of the quantum wave tells us the probability for something to happen. If the quantum wave does not change, this means that the probability for this or that outcome does not change. So, for example, if today there is a 25% chance of space bending in a certain way, tomorrow there will be the same 25% chance of space bending in the same way. And the day after, and the day after that, on and on forever. Does that make sense?

Actually that makes no sense at all. Imagine a weather forecaster telling us that tomorrow there is a 25% chance of rain. And the day after there is also a 25% chance of rain. And the day after that, on and on forever. How can the chance of rain be the same every day? That could happen only if the earth's atmosphere was somehow fixed, static, or frozen, with nothing changing or moving. It seems that, at the deepest level of physics including gravity, nothing is changing, nothing is moving, and nothing is happening. At bottom, the universe is frozen and timeless. Or at least, that is what the Wheeler–DeWitt equation seems to imply.

There is a long history of attempts to take this startling conclusion seriously, beginning notably with DeWitt himself in his classic paper on quantum gravity published in 1967. DeWitt claimed that time 'emerges' only as an approximation. At the fundamental level, as described by the Wheeler–DeWitt equation, time does not really exist. When we look at how Einstein's curved space interacts with ordinary matter, an *appearance* of time emerges from the equations. If that sounds obscure, you are in good company.

To get an idea of how this is supposed to work, think of a car driving along a winding road. Normally we would say that, as time passes, the car moves from one place to another. But in a timeless description, we might simply have a certain route taken by the car—effectively a line drawn on a map, showing the car's journey—with nothing said about time or about how fast or slow the journey was (Figure 72). In a quantum description we might be given the probability for this or that route. But let us just say there is a route drawn on a map, which represents the journey taken by the car. If we look more closely, we notice that the amount of fuel inside the car is not the same all the way along the line. At one end there is more fuel. We might label this 'start'. At the other end there is less fuel. We might label this 'finish'. And in between, from one end of the route to the other, the amount of fuel diminishes. Normally we would say that, well, obviously, as time passes and the car moves, it consumes fuel, so the amount of fuel in the tank goes down. But now we are trying to think about the car with no concept of time at all. How can we do that? Here is an idea. We might use the amount of fuel consumed as an effective measure of 'time'. The car will certainly have a fuel gauge indicating how much fuel is left in the tank. So we are saying that the fuel gauge might be deployed as a sort of clock. If we want to know what time it is, we can look at the fuel gauge. The less fuel left in the tank, the more time has passed. How does that sound?

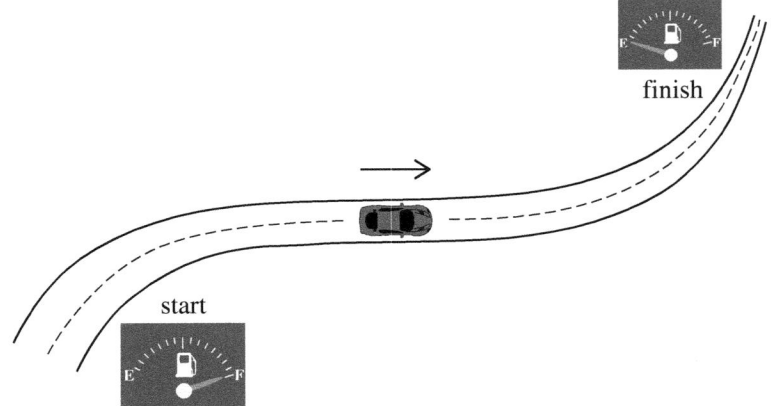

Figure 72. Understanding a car journey without time. The tank of fuel starts full and finishes empty. The fuel gauge can serve as a clock.

Well, we might begin to wonder. What happens when the car runs out of fuel? Obviously, the car will stop moving. But does that mean time itself comes to a standstill? And what if we fill up the tank? Does that mean time somehow flows backwards as the tank fills up? Perhaps this is not such a good idea after all.

All this may sound rather silly. But similar ideas have been pursued in quantum gravity for more than half a century. For example, literally hundreds of papers have been published trying to apply the Wheeler–DeWitt equation to understand cosmology and the early universe. We saw in Chapter 6 that our universe started with a hot Big Bang and has been expanding ever since. Quantum gravity is likely to be important near the beginning, when the universe was very small and very dense and expanding rapidly. But there is a problem. According to quantum gravity, at least as usually understood, the quantum wave for the universe is static and frozen. And yet we see the universe changing and expanding. How can we understand that? Here is a common trick. Let us use the size of the universe as an effective measure of 'time'. When the universe is smaller, that means 'earlier'. When the universe is larger, that means 'later'. So the size of the universe acts as a clock. If we want to know what time it is, we can look at the size of the universe. How does that sound?

Well, again, we might begin to wonder. What happens if the universe expands to a certain maximum size, stops, and then recontracts?

(That could happen if the universe contains enough matter, so that gravity is strong enough to pull everything back together again.) Would this mean that time flows to a certain maximum point, stops, and then runs backwards?

Let us pause and think for a moment about why we do not have this kind of trouble in everyday life. Perhaps we have a clock whose battery has run out. The clock has stopped. But we do not conclude that time itself has stopped. That is because we believe there is something called time that exists even when there is no clock there to measure it. The clock does not really 'define' time; it merely measures it. This is a tricky area to discuss properly. But most people will understand that a well-designed and correctly functioning clock (or wristwatch) will measure time accurately, while a half-broken clock will not. If someone pushes the hands of a clock to make them go in reverse, no one will think that time is going backwards. But all this assumes there really is such a thing as time, which exists independently of any equipment that may be employed to measure it. If instead we try to understand physics without time, we end up with the sort of contortions sketched above. We look for something concrete to play the role of time—fuel in the tank, or the size of the universe—but there is no guarantee that something will behave the way it should.

Since the work of Wheeler and DeWitt in the 1960s, the meaning of time in quantum gravity has remained contentious. There are those who suggest that something is wrong or missing. Perhaps we need to reinstate time into the theory somehow. This is the view taken, for example, by the radical American theorist Lee Smolin, as well as by this author. Others insist that we accept a fundamental physics without time—or at least, without time as we usually understand it. Contemporary champions of the latter view include the British physicist Julian Barbour and the Italian physicist Carlo Rovelli (even if their ideas differ in certain respects).

There could hardly be a deeper or more far-reaching controversy in science. To convey a sense of what is at stake here, let us consider the fossil record (Figure 73). When we go digging in the ground we find the fossilised remains of animals and plants long since dead. When we dig deeper we find different kinds of fossils. The deeper we go, the more primitive they become. We understand that the deeper we dig, the older the fossils and the further we are looking back in time. Each layer of rock contains, in effect, a snapshot of the living biology of that time. So the successive layers are like successive stills from a film reel, telling the story of evolution

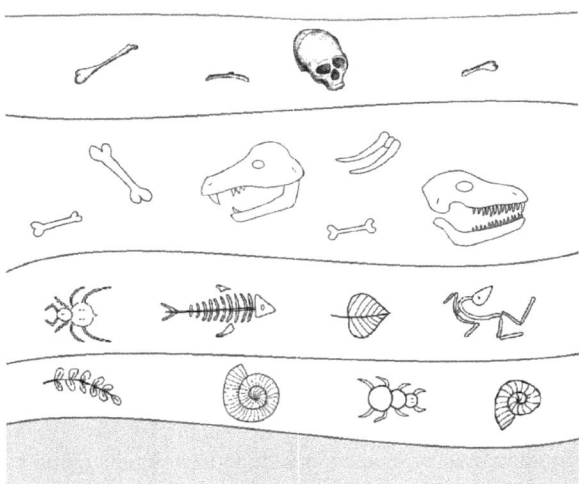

Figure 73. Is 'the past' merely present memory? Geologists and evolutionary biologists tell us that each layer (or stratum) of rock records the living biology of the remote past. The deeper the layer, the older the record. And yet this fossil record might be said to exist only in the present.

from primitive to more advanced life forms.[62] Note that the fossil record is static: it is not moving or changing (at least not over human timescales); it is just there in the layers of rock. But it has what we might call a structure or pattern, from which we can deduce the past history of the evolution of life on Earth (for example, the sudden extinction of the dinosaurs some 66 million years ago). Note also that, in geology and evolutionary biology, we usually assume that past history really happened, even if it was millions of years ago—just as last year really happened, yesterday really happened, and five minutes ago really happened. The past is not merely a theoretical deduction or mathematical fiction, nor is it simply a memory in our heads or in rocks. It really happened.

What does all this have to do with quantum gravity? Well, like the fossil record, the quantum wave (as given by the Wheeler–DeWitt equation) is static and unchanging. It is just there, like those layers of rock. But it has a structure or pattern, which can give the *appearance* of a past history. Barbour has emphasised that, in his view, while we can see something

---

[62] We are simplifying a bit. Because of continental drift (plate tectonics) and other geological effects, layers that were further below can sometimes be pushed further up.

analogous to a past history etched inside the mathematical structure of the quantum wave, the universe is really timeless. Nothing happens. There is no such thing as time. Our universe may *look as if* it has a past history, but that is an illusion. This is in some respects reminiscent of the view taken by 'young Earth creationists'—that the fossil record was created intact, by a supernatural event taking place about six thousand years ago, with a structure that gives the appearance of a much longer past history. Despite this similarity, however, Barbour and others of this view do not regard themselves as creationists. Like most scientists, they accept the usual views of mainstream evolutionary biology. And yet, in the peculiar context of quantum gravity, they find themselves arguing that, somehow, the past is an illusion and that in general the passing of time is an illusion.

In contrast, in his 2019 book *The Order of Time*, Rovelli argues that time does exist even at the deepest level, but not as we usually know it. On Rovelli's view, we need to accept a less orderly notion of time, grounded in the haphazard relations between things. Our usual orderly idea of time emerges only in certain approximate conditions.

It is worth emphasising that these arguments about time in quantum gravity are not mere semantics or armchair philosophising. They are part of the nuts and bolts of trying to understand the early universe by means of quantum mechanics combined with Einstein's theory of gravity. For example, in the 1980s numerous papers were published by Hawking and his school, in which they tried to calculate the probability for our universe to be large or small (as well as the probability for it to be curved or flat). This was done with the static, unchanging quantum wave as derived from the Wheeler–DeWitt equation. Hawking believed that we could apply Born's usual formula. So he assumed that the probability for this or that was equal to the squared magnitude of the quantum wave. From this some results were derived concerning the probable size of our universe. There was a seductive simplicity and directness to Hawking's approach. But in the end it did not make sense. The calculated probabilities do not change with time. They are like the weather forecast predicting a 25% chance of rain today, tomorrow, and always. There was no way to make sense of a changing or evolving universe. Hawking had even suggested that, if the universe expanded to a maximum size and recontracted, the flow of time would reverse. Unsurprisingly, this idea turned out to be mathematically flawed (as shown by Canadian physicists Don Page and Raymond Laflamme). Hawking's approach has since been widely abandoned. It is now known, somewhat disparagingly, as the 'naïve' interpretation of quantum gravity. Tempting as it may be, we cannot appeal to Born's formula to calculate probability from the quantum wave.

There are more sophisticated approaches that attempt to calculate probability in some other way from the timeless quantum wave. Somehow the theory has to be able to explain at least the appearance of time passing. Despite more than half a century of effort, however, there are still difficulties and controversies, which are usually lumped together under the general heading 'the problem of time'. We can hardly do justice in a few paragraphs to this difficult and complex area. Some authors believe there is no problem. When confronted with seemingly paradoxical results, they point out that our received notions of how time behaves might not be valid in all circumstances, and that our ordinary notions of time might emerge only in special conditions. Others believe something is wrong or missing.

In 2017 the British physicist Edward Anderson published his book *The Problem of Time*, in which he gave an exhaustive and highly technical review of the field. The book runs to nearly a thousand pages and draws no definite conclusions. It seems fair to say that, after more than half a century since Wheeler and DeWitt wrote down their famous equation, the problem of time is still with us.

Pilot-wave theory to the rescue?

According to quantum mechanics, there is a quantum wave and nothing more. So if the wave is fixed and unchanging, it seems that the universe is static and timeless. But in pilot-wave theory the quantum wave is not everything. There are also particles moving around in space. There are also fields oscillating and shifting. If we include gravity, space is bending and stretching. And all of this movement and change is guided by the quantum wave or 'pilot wave'. In effect, the quantum wave tells things how to move—regardless of whether those things are particles, fields, or curved space.

So we should be able to write down a pilot-wave theory of quantum gravity. We already have an equation for the quantum wave (given to us by Wheeler and DeWitt). We just need something like de Broglie's law of motion for particles, except now it will be a law of motion for a curved space. This can in fact be done. And it has been done. The basic equations were first written down in the 1990s, by several authors, most notably and in detail by the Japanese physicist Tsutomu Horiguchi. Mathematically the theory is similar to the pilot-wave theory of particles and fields. The main difference is that the quantum wave is static. But this does not mean the curved space itself is static. Even though the guiding wave is fixed, it pushes the system around and so the curvature of space changes with time. The universe is no longer timeless. Space is bending and stretching (or shrinking) as time passes, as it should (Figure 74).

This theory has been developed in particular by the Brazilian cosmologist Nelson Pinto-Neto and co-workers, whose main interest has been to understand how the Big Bang could have originated from a 'bouncing cosmology'. The idea is that, before the Big Bang, the universe was large and contracting. It shrank down to a tiny size and then underwent a bounce, whereupon it started to expand to give the universe we see today. Related ideas have been considered by other workers, for example the 'ekpyrotic universe' model developed by American cosmologist Paul Steinhardt, South African cosmologist Neil Turok, and collaborators. These models are often regarded as potential alternatives to inflationary cosmology.

What is most distinctive about the work of Pinto-Neto and his school is the application of pilot-wave theory to understand the details of the cosmological bounce. Remember that quantum mechanics only tells us about the probabilities for outcomes of measurements. If there is no observer present

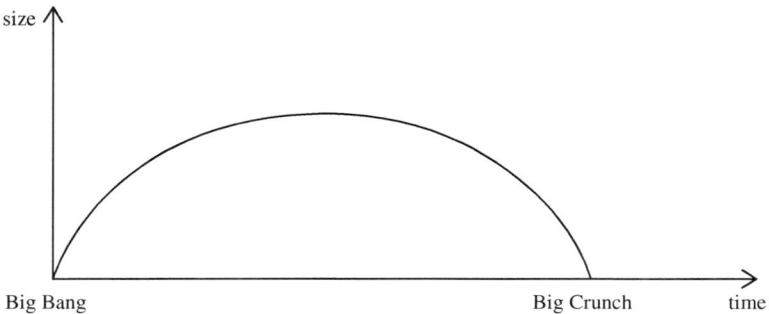

Figure 74. Example of an expanding and recontracting universe. As time passes, from an initial Big Bang, the size of the universe increases to a maximum and then decreases, ending in a 'Big Crunch'. (Technically, the size is measured by what cosmologists call the 'scale factor', which is a measure of overall linear scale.)

to make a measurement, the theory does not really say anything concrete. This means the theory is embarrassingly silent about what was happening in the early universe (for example, during a bounce), when of course there were no observers making measurements. In contrast, in pilot-wave theory, in addition to the quantum wave there is also an evolving space, with moving particles and fields, and all this can be discussed in the absence of observers. We can, for example, calculate what happens to space during a bounce, provided we are able to solve the equations (for a suitably simplified model).

In the past twenty-five years Pinto-Neto and collaborators have produced a large body of work, showing that pilot-wave theory has many advantages when it comes to 'quantum cosmology' (the application of quantum mechanics to the whole universe including gravity). Even so, most workers in the field continue to ignore pilot-wave theory and try to make do with standard quantum mechanics. The results are often conceptually painful. On the one hand we have the quantum measurement problem, where nothing seems to exist without an observer there to see it. On the other hand we have the problem of time, where the universe including gravity seems to be in a state of suspended animation with nothing happening. Put the two together and theorists find themselves struggling to say anything clear or sensible. Despite the work of Pinto-Neto and his school, the field of quantum gravity and quantum cosmology remains gripped by the timeless approach and handicapped by the devastating combination of the problem of measurement and the problem of time. After half a century of struggling with such a multitude of confusions, perhaps it is time to rethink the whole subject in terms of pilot-wave theory.

The problem of probability

We have seen that in quantum gravity there are problems when we try to calculate probabilities from the timeless quantum wave. In fact the 'problem of probability' is even worse than we have been letting on. And a similar problem appears in pilot-wave theory as developed by Horiguchi, Pinto-Neto, and others.

Here is the full scale of the problem. When we apply probability theory, the probabilities for all possible outcomes must add up to 100%. So, for example, if we toss a large number of (biased) coins, maybe 40% of them will show heads while 60% of them will show tails. Obviously, the total must add up to 100%. In fractional terms the probabilities—in this example 0.4 and 0.6—must add up to 1. But in quantum gravity, when we add up the total probability, the answer is usually *infinite*.

What is going on? The problem comes from the peculiar structure of the Wheeler–DeWitt equation. When we solve it we find a static quantum wave that is so spread out it covers all possibilities without diminishing.[63] To see what we mean, think of a wave on a pond, spreading out from where a stone was thrown in. The wave covers only a small region of the surface of the pond. If the (squared) height of the wave is a measure of probability, adding up the total probability over that small region gives a finite answer. But imagine if instead the pond extends to infinity—and the wave is spread all over that infinite space, without ever getting smaller (Figure 75). Adding up the total probability then gives an infinite result. In quantum gravity, something similar happens with the quantum wave.

So if we try to calculate probabilities from the squared magnitude of the quantum wave (as done by Hawking) we run into a severe problem. Not only are the probabilities static (which is bad enough). When we add them up the answer is infinite. Technically, the probability distribution is 'non-normalisable', which means that it is not a properly defined probability. This problem has plagued the theory of quantum gravity for decades, and of course many ingenious schemes have been devised to try to avoid it.

But now let us return to quantum gravity as described by pilot-wave theory. As we have discussed at length in this book, according to pilot-wave theory, when a system reaches the state of quantum death the probability for it to be found here or there will be given by Born's quantum formula. We might expect something similar to happen in pilot-wave quantum gravity. When the system reaches quantum death, Born's formula should tell us the

Figure 75. A world of infinite probability. If the wave is spread over infinite space (here shown in one dimension) without ever getting smaller, and if the squared height of the wave is a measure of probability, then the total probability adds up to infinity.

[63] Technically, the Wheeler–DeWitt equation has the structure of a Klein–Gordon equation on configuration space, yielding solutions that are non-normalisable.

probability of finding this or that curved space. But Born's formula says that probabilities are to be calculated from the squared magnitude of the quantum wave. And we have just seen that if we try to do that in quantum gravity, the total probability adds up to infinity. So it seems that, in the state of quantum death, we will end up with the same problem of infinite total probability—which makes no sense at all.

For this reason workers in pilot-wave quantum gravity have shied away from discussing probability. Much work has been done, for example, calculating the possible trajectories of the early expanding universe (as in Figure 74). But if we try to assign a probability to this or that motion, we run into trouble. Bizarrely, we find ourselves in a position opposite to that of quantum mechanics: we can calculate detailed trajectories, but we cannot calculate probabilities.

This leaves us with a logical gap. At some point we will need to describe particles (or fields) moving on a background expanding space. We will need to recover the usual approximate theory, where the particles have a quantum wave that depends on time. In the state of quantum death those particles should obey Born's formula, as experiments today confirm. In principle we should be able to describe all of this with the underlying theory of quantum gravity. But we have seen that in the underlying theory we are unable to define or calculate probability. So where will the probability for our particles come from? Put differently: how can we explain the probabilities we see today for particles moving in space, when according to quantum gravity at the deepest level there is no properly defined probability?

We seem to be in trouble. As we have said, workers in the field have shied away from this issue. But until it is addressed, ultimately even the pilot-wave theory of gravity makes no sense.

Quantum death transcended

There is a way forward—if we are willing to take a radical step. We must accept that, in quantum gravity, Born's formula is always wrong. Probability is never equal to the squared magnitude of the quantum wave. Remember that if we do apply Born's formula to calculate probabilities, the total adds up nonsensically to infinity, whereas proper physical probabilities must always add up to 100%. So Born's formula must be wrong at all times. That is our radical solution to the problem of probability.

To put this idea in perspective, note that there has been a long history of trying to modify the quantum wave so as to eliminate the infinity. When we do that, however, we encounter other problems. And this kind of fiddling has been going on for more than half a century. Instead, why not accept that in quantum gravity Born's formula is wrong? That formula just does not apply to gravity. There is a quantum wave, which in pilot-wave theory guides the motion of the bending and stretching space (as well as the motion of particles and fields on that space). But the probabilities can never match the squared magnitude of the quantum wave. In other words, Born's formula can never be correct. At the deepest level, we need to forget about Born's formula.

At this point the reader might object. What about quantum relaxation? In previous chapters we saw examples where the probabilities start off differently from those of Born's formula, but nevertheless end up the same. Quantum relaxation pushes us towards the state of quantum equilibrium, or quantum death, where Born's formula is correct. Why can that not happen in quantum gravity? It cannot happen because it is mathematically impossible. If the probabilities did end up matching the squared magnitude of the quantum wave, they would need to add up to infinity, which simply cannot be for a properly defined probability. And so quantum relaxation cannot occur.[64]

So now it is getting interesting. We are arguing that, in quantum gravity, *quantum relaxation is impossible*. No matter how much time passes, we can never reach the state of quantum equilibrium. This means that, at the deepest level of gravitational physics, there is no state of quantum death. Truly, gravity can save us.

If we accept this idea, the problem of probability disappears. The true probabilities will always add up to 100% as they should. But those probabilities can never match Born's formula.

Now we have to think through the implications. In quantum gravity, Born's formula is always wrong. Quantum nonequilibrium lasts forever. No matter how long we wait, we will never reach quantum death. And yet we see quantum death all around us. How?

How the universe dies

There is a simple explanation. At the very earliest times in our universe, the energies were so high that quantum gravity was important. To describe

---

[64] This has recently been illustrated by computer simulations carried out with the Indian physicist Adithya Kandhadai.

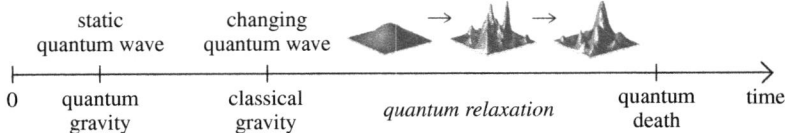

Figure 76. How the universe dies. In the very early quantum-gravity era there is no quantum relaxation. As the universe expands and cools, eventually quantum gravity can be neglected and quantum relaxation can begin, leading to the quantum death we see all around us today.

that era, we have no choice but to apply the Wheeler–DeWitt equation and a static quantum wave. As we have seen, in those conditions there can be no quantum death. A little later, however, as the energies decrease we can ignore the effects of quantum gravity and define a changing quantum wave for matter, which obeys the usual Schrödinger equation. We are then back to the usual quantum wave. There is now no problem of probability and quantum relaxation can begin as usual. And once quantum relaxation begins, the rest is history. Soon we end up in our present state of quantum death, where everything around us is imbued with the same quantum noise, where the uncertainty principle reigns, and where we are unable to see or control the finer details of quantum reality. As far as we can understand, that is how the universe 'dies' (Figure 76).

Let us look at this again in a bit more detail. In quantum gravity we apply quantum mechanics not only to the matter and fields that are moving around in space but also to space itself. We find that the quantum wave is static, and quantum probabilities seem to add up (paradoxically) to infinity. But quantum gravity is important only at extremely high energies, in the very earliest moments of our universe. At slightly lower energies, we expect the usual quantum wave to emerge as an approximate description. We can then apply quantum mechanics only to the matter and fields that are moving around in space. Space itself is treated as a given background described by Einstein's classical theory of gravity. Quantum-gravity theorists call this the 'semiclassical approximation', because space is treated with classical physics while matter is treated with quantum physics—a sort of halfway house between classical and quantum. With that approximation in place, we are back to the usual pilot-wave theory of particles and fields on a background space, which we have discussed in previous chapters. And so we can expect rapid relaxation to quantum death.

According to our interpretation of quantum gravity, then, in the very early universe Born's formula is permanently broken—quantum relaxation cannot happen, and quantum death cannot be reached. The very early universe is necessarily in a state of quantum nonequilibrium, and in this sense remains very much 'alive'. A bit later, though, when the universe has expanded somewhat and the energies are lower, we can neglect the effects of quantum gravity and apply the usual changing quantum wave to describe particles and fields in the background expanding space. At that point quantum relaxation starts and we quickly reach the state of quantum equilibrium or quantum death in which we now find ourselves.

A glimmer of light?

At this point the reader might find the whole discussion as depressing as before. Well, alright, that is certainly interesting that we can avoid quantum death when quantum gravity is important. But that is the case only in the very early universe. As the universe expands we soon leave what physicists call the 'deep quantum-gravity regime'. In other words, quantum-gravity effects soon become small to the point of being more or less negligible. Once that happens, quantum relaxation can begin, and we quickly end up in the state of quantum death (Figure 76). So gravity does not really save us after all. We still find ourselves in the same place, trapped in quantum death forever and ever.

It might seem that way. But remember that the usual Schrödinger equation, with a changing quantum wave, is only approximately true. Even when distances are large and energies are small, there will be *some* effect from quantum gravity. The effects will be tiny, but they will not be exactly zero. So now we have the germ of an idea. Let us look more closely at the Schrödinger equation, remembering that according to quantum gravity it ultimately comes from the Wheeler–DeWitt equation calculated in a certain approximation. What if we rerun the calculation more precisely? We might expect to find something that is like the Schrödinger equation but *a little bit different*. In the jargon, we might expect to find 'quantum-gravitational corrections to the Schrödinger equation'. In that case, we might begin to wonder: might those corrections help us somehow escape from quantum death? After all, if quantum death can occur only when we ignore quantum gravity, perhaps if we include even small effects from quantum gravity, there might be some way to escape. Sounds plausible?

This intuition turns out to be justified. Not only will we find a promising way out of quantum death, we will also resolve a mystery that has been hanging over the field of quantum gravity for the past thirty years.

The curious case of impossible probability

The year is 1991. German physicist Claus Kiefer and Indian physicist Tejinder Singh have just published a paper in which they rethink how the Schrödinger equation (for a changing quantum wave) emerges from the deeper Wheeler–DeWitt equation (for a static or timeless quantum wave). They rework the mathematical steps thoroughly, showing with admirable clarity how to obtain the Schrödinger equation as an approximation. Crucially, they then go further. They are able to find extra terms that amend the Schrödinger equation very slightly. These extra terms represent tiny corrections caused by quantum gravity. For example, if we fire an electron at a two-slit barrier, the changing quantum wave will obey the usual Schrödinger equation to a very good approximation. But there will also be tiny correction terms, as calculated by Kiefer and Singh. Because of those terms, the quantum wave will be very slightly different from usual. Under ordinary laboratory conditions the difference is far too small to measure, but in principle it is there and might be measurable one day.

But now there is a puzzle. Some of the correction terms have a clear physical meaning within quantum mechanics. These terms slightly distort the quantum wave, and produce a tiny shift in atomic energy levels (far too small for us to measure but there in principle). This is the sort of thing that would be expected. There is, however, something else. Kiefer and Singh find another class of correction terms that destroy the internal consistency of the standard theory. Remember we said that bona fide probabilities must add up to 100% (when summed over all possible outcomes). There is actually a bit more to say. If the probabilities add up to 100% now, they must *also* add up to 100% later. The total probability must stay the same, now, tomorrow, and always. This is known as the 'conservation of probability'. And quantum mechanics usually guarantees that this is true. The quantum wave changes with time, but the total probability does not. And this had better be so, for otherwise quantum mechanics would collapse in an inconsistent and nonsensical heap. This is where Kiefer and Singh come in. Some of their correction terms destroy the conservation of probability

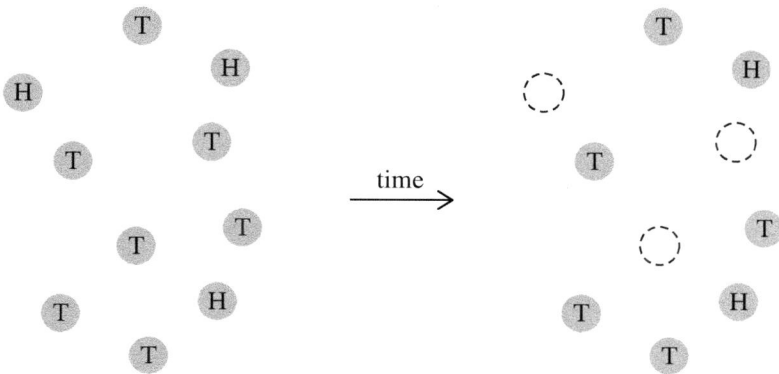

Figure 77.   An impossible disappearing probability. Coins that were showing 30% heads and 70% tails turn into coins that are showing 20% heads and 50% tails. What about the other 30%?

for the quantum wave.[65] The total probability *changes* as time goes on. Or at least: if we insist on using Born's formula to calculate probabilities, we find that the total probability changes. And that is impossible.

Kiefer and Singh were rather nonplussed by this result. If the total probability changes over time, we cannot make physical sense of the equations. Imagine if, for example, today we toss a (biased) coin many times and we find 30% heads and 70% tails. Then tomorrow we take the same coin, toss it again many times, and now we find 20% heads and 50% tails (Figure 77). Excuse me? What about the other 30%? Exactly. It does not make sense. The total probability must stay at 100%, today, tomorrow, and always.

Kiefer is a highly respected authority on the theory of quantum gravity and the author of a well-known book on the subject. Over the past thirty years he and his collaborators have repeatedly reworked the equations in various contexts, including for models of inflationary cosmology. The 'impossible' terms keep coming up. To handle these terms Kiefer and collaborators adopt a simple but effective strategy: ignore them. Keep only the terms that make sense and apply these to calculate new effects from quantum gravity (for example, very small corrections to the CMB). The impossible terms make no sense and are set aside. This strategy

---

[65] Technically, these corrections amount to tiny 'non-Hermitian' terms in the Hamiltonian operator, resulting in a violation of 'unitarity'.

is of course only a stopgap, as Kiefer and collaborators well know. The impossible terms are still there. No one knows why. Might those terms be telling us something?[66]

Quantum disintegration: or final escape from quantum death

The impossible terms make no sense—*if* we insist on using Born's formula to calculate probability. But they do make sense in pilot-wave theory. Those terms are telling us that, under the right conditions, we can finally escape from quantum death.

To see this, imagine we have a particle that has already reached quantum death. This means that its probability is equal to the squared magnitude of the quantum wave (Born's formula). Normally, once a system reaches quantum death, it stays there forever. If Born's formula is true today, it will be true tomorrow and always. But now we have the impossible terms to contend with. To make sense of them, we need to bear in mind that in pilot-wave theory the true probability can be different from the usual quantum 'probability' calculated from Born's formula. Here is how it works. We know that the true probability for our particle will always add up to a total of 100%. We also know that the total 'probability' calculated from Born's formula can change from 100% to something else (because of the impossible terms). And so let us bite the bullet and make a deduction. As time passes, the true probability must drift away from the (false) 'probability' as calculated from Born's formula—since the former remains at a total of 100% while the latter does not. In other words: if the particle starts off obeying Born's formula, later the particle must violate Born's formula. There is no paradox of impossible probability. Instead it is simply the case that, if the particle starts off in a state of quantum death, later the particle will *escape* from quantum death (Figure 78).

We seem to have arrived. The impossible terms that were first discovered by Kiefer and Singh in 1991 make no sense in ordinary quantum mechanics—where the textbooks insist on using Born's formula to calculate probability. But those terms make perfect sense in pilot-wave theory, where the true probability can be different from usual. Those small

---

[66] Some workers have suggested alternative methods of deriving the semiclassical approximation, in which the impossible terms do not appear. However, these alternatives may turn out to be mathematical subterfuges that disguise real physical effects. At the time of writing, it is not known for certain which method correctly describes the time-dependent quantum waves found in nature.

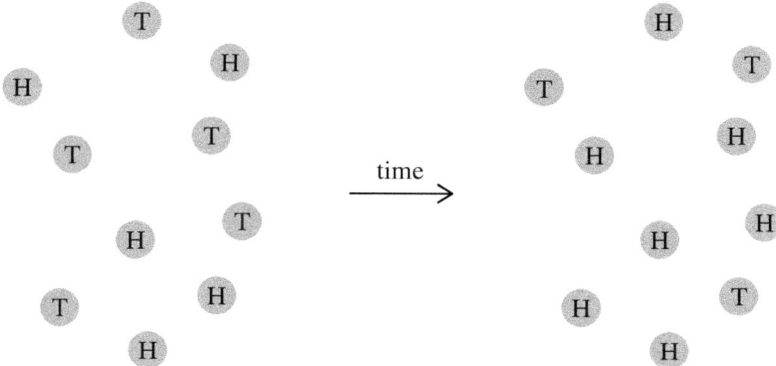

Figure 78. Escaping quantum death. In our analogy with coins, a 50:50 mix of heads and tails changes into a 70:30 mix, and so moves away from equilibrium. Something similar can happen in pilot-wave theory when quantum-gravitational corrections are included: particles that initially obey Born's formula can end up breaking it.

terms, induced by gravitational effects, drive a system away from quantum death. Born's formula is 'unstable': if it works today, it need not work tomorrow. In a word, gravity can overcome quantum death.

Let us pause for a moment and take in the significance of where we are. For over thirty years the strange terms found by Kiefer and co-workers have been regarded as an awkward artifact to be ignored. We are suggesting that those terms can show us a way out of quantum death, by *creating* a state of quantum nonequilibrium or quantum life. At the beginning of the twentieth century we learned that matter can disintegrate through radioactive decay. According to our reasoning, quantum mechanics itself can disintegrate—if we wait long enough.

How long will it take? By including the 'impossible' terms in the equations of pilot-wave theory, we can calculate how long it takes for a system to drift away from quantum death. This process of quantum disintegration, or quantum instability, has been studied theoretically for various cases.

For example, in inflationary cosmology, quantum fluctuations of the early inflaton field are imprinted on the CMB (Chapter 6). We can now study the effect of quantum disintegration on the inflaton field. Specifically, we can find out how far the early inflaton field might drift away from quantum death, resulting in anomalous statistics in the CMB sky today. Unfortunately, however, according to the approximate calculations done so far, the effect will be so tiny as to be unmeasurable. Even so, the example

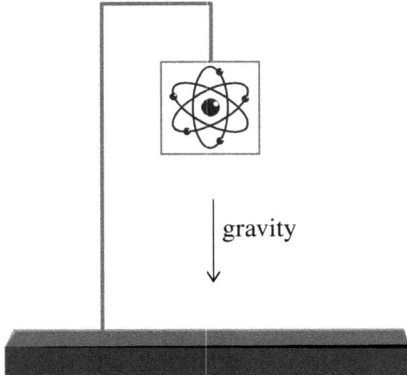

gravity

Figure 79. Waiting for nonequilibrium. Atoms suspended in the earth's gravitational field could in principle drift away from quantum death and deviate from Born's formula.

is instructive. It shows how, in a realistic setting, a system can drift away from quantum death and build up violations of Born's formula—taking it outside the domain of accepted physics.

If gravity is able to overcome quantum death and produce violations of quantum mechanics, might we see this in the lab? What if we trap quantum systems in a box, suspended in a gravitational field, and wait (Figure 79). Will they drift away from equilibrium and the usual laws of quantum physics? In principle, the answer is yes. For example, atoms in the earth's gravitational field could drift away from quantum death and deviate from Born's formula. But there are two hard caveats. First, even under ideal conditions, it would take many ages of the universe for the effect to build up significantly. In principle the effect should be there, but far too tiny to be measurable in our lifetimes. Second, in practice, we will have quantum relaxation to contend with. As nonequilibrium deviations from Born's formula build up over time, quantum relaxation could quickly erase them again. There will be a net drift away from quantum death only in conditions where relaxation can be neglected. Unfortunately, for technical reasons, it is difficult to see how such conditions could be achieved in this case.[67] Perhaps there is some way around this, but at the time of writing

---

[67] According to the calculations, there will be quantum instability for atoms in a gravitational field only when their Hamiltonian is changing with time, making it particularly difficult to avoid relaxation.

just waiting patiently in a gravitational field does not look like a remotely practical way to escape from quantum death.

At this point we might again have that sinking feeling. Is there no realistic, measurable way out of quantum death? There is, in fact, at least one situation where the effects of quantum disintegration can be significant, and that involves the mysterious objects known as black holes.

We touched on the topic of black holes in Chapter 7. We saw that very small 'primordial' black holes are likely to have formed in the early universe. They may well populate the universe today. And they could be a significant, or even dominant, component of dark matter. As shown by Hawking, owing to quantum effects black holes can radiate energy and eventually explode. Some astronomers are searching for the telltale signs of such explosions in our galaxy and beyond. So now it is interesting. We can look at how the impossible terms found by Kiefer and co-workers (reinterpreted with pilot-wave theory) affect the radiation coming out of a black hole. As we might expect, the emitted radiation is pushed slightly away from quantum death and from Born's formula. So, for example, if the photons emerging from a black hole were to hit a two-slit barrier, we would not see precisely the usual wavy interference pattern (Figure 80). Born's formula would fail to some extent and quantum mechanics would be broken.

Crucially, though, the effect of the impossible terms will be significant only in the final stages, when the black hole has already radiated away most of its mass. According to the approximate calculations done so far, an evaporating black hole ends its life in a final burst of radiation that transcends quantum death and violates the laws of quantum physics. Because the effect is significant only in the final burst, the predicted failure of Born's formula will be difficult to detect in the sum total of the radiation we receive. But, in principle, at some level of accuracy Born's formula will fail.

The *Fermi Gamma-ray Space Telescope* is currently scanning space for telltale gamma rays that might be produced by disintegrating dark matter or by exploding primordial black holes (Chapter 7). In either case, according to the arguments given in this book, if such rays were found it would be important to test them for possible violations of quantum mechanics. This might involve something like a two-slit interference experiment, or an experiment that tests polarisation probabilities. Such experiments would, however, have to take place aboard a dedicated satellite in space, since otherwise the incoming photons will be absorbed by the earth's atmosphere.

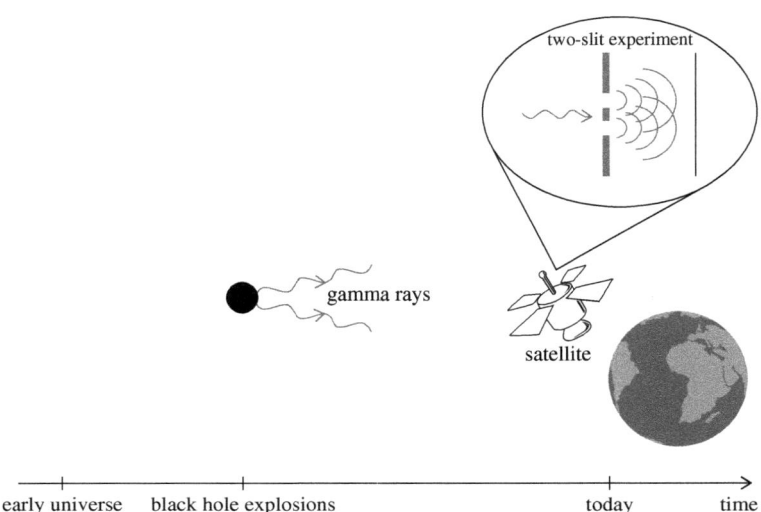

early universe   black hole explosions   today   time

Figure 80. Primordial black holes, created in the early universe, eventually evaporate and explode, emitting radiation (including gamma rays). According to pilot-wave theory, exotic effects from quantum gravity can drive the emitted radiation away from quantum death, resulting in small violations of Born's formula.

Should violations of quantum mechanics be discovered, the scientific and technological implications would be huge (Chapter 10). No doubt techniques would be developed to harvest this new resource and make it available on Earth—perhaps by using the incoming gamma rays to knock ordinary particles out of equilibrium, on the satellite, before sending them back to Earth.

We conclude that radiation from exploding primordial black holes could enable our escape from quantum fog. Should astronomers detect such radiation, we might have what we are looking for: a supply of 'subquantum' photons that evade the obscuring effects of quantum noise. Such photons could be deployed to beat the uncertainty principle—opening up quantum reality to proper scrutiny and control.

We may have finally found our way out of quantum death. To do this we had to rethink some of the deepest questions in quantum gravity. When properly understood, quantum gravity shows us that quantum death may not be forever. At the end of this long journey, our hopes are pinned on the most enigmatic of objects: evaporating black holes. It is now time to look more closely at black holes and at the role they may play in our universe.

9
Black Holes and the Edge of Physics

Black hole explosions

There is much that is not understood about black holes. In brief, a black hole forms when a sufficiently large mass becomes concentrated within a small region of space. This might happen, for example, if a star runs out of nuclear fuel and begins to collapse under its own weight. As the star contracts and gets smaller, at a certain point the gravitational pull at the surface can become so strong that not even light can escape (Figure 81). That surface is called an 'event horizon'. It marks the point beyond which the collapsing material is unable to communicate with the outside world. The star becomes totally dark and incommunicado. An external observer can no longer see what is happening. Where there was once a collapsing star, now there is only a 'black hole'.

Within that small space the mass of the star continues to collapse under its own gravity. According to Einstein's theory, this process of 'gravitational collapse' continues until, in effect, all the original mass of the star is concentrated at a point of infinite density called a 'singularity'. But Einstein's theory neglects quantum effects. There is reason to believe that, as the singularity is approached, at some point quantum gravity will become important and a literally infinite density will be avoided. In any case, as far as the outside world is concerned, precisely what goes on close to the singularity is forever unknowable. The interior of the black hole is shielded from view by the event horizon: no light can escape, and no signal or message or communication of any kind is possible from the interior to the exterior, not now and not ever.

Or at least, that is what the textbooks tell us. The idea of an event horizon is based on the assumption that no signal can travel faster than the speed of light. If light can no longer escape from the surface of a collapsing

Beyond the Quantum. Antony Valentini, Oxford University Press. © Antony Valentini (2025).
DOI: 10.1093/oso/9780198853749.003.0010

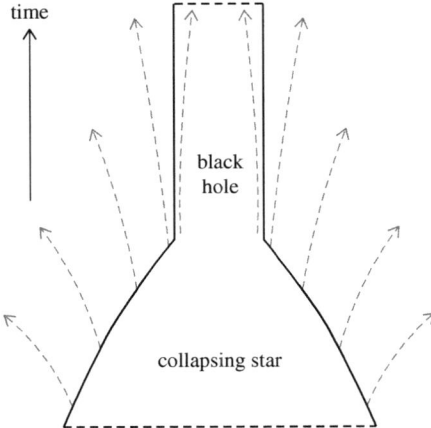

Figure 81. A star collapses under its own weight to form a black hole, from which not even light (dashed lines) can escape. (Technically, the collapse is depicted in terms of 'ingoing Eddington–Finkelstein coordinates'.)

star, then communication with the outside world is cut off forever. Unless the assumption is wrong and superluminal signalling is possible after all.

We have seen that superluminal signalling is, in fact, possible if we have entangled particles that violate Born's formula (Chapter 5). So if some of the atoms in the collapsing star were entangled with atoms in the exterior region, and if those atoms had somehow evaded quantum death, then there would be no true event horizon: signals could pass from the inside to the outside.[68] In other words, if we could escape from quantum death, we would be able to peer behind the so-called event horizon and find out what really happens to the collapsing star.

But let us backtrack a little and consider how things look in the state of quantum death where quantum mechanics can be applied. As we mentioned at the end of Chapter 7, Hawking understood that, when we take quantum mechanics into account, a black hole is not completely black but radiates energy like a hot body. As a result the black hole steadily loses mass and becomes smaller. The smaller the hole becomes, the higher the

---

[68] The signals would travel along a preferred hypersurface or 'slicing' of spacetime (associated with an absolute time).

temperature, and the faster the hole loses mass. This process can spiral out of control, resulting in an explosion. How long does it take? For a black hole formed by stellar collapse, the mass is so large and the temperature so small that it would be impossible for us to notice anything in practice—it would take many times the age of the universe for the mass of the hole to be significantly depleted. But for small, primordial black holes, the whole process takes much less time and there is every reason to believe that such objects could be exploding in our universe right now—hence the search for telltale gamma rays by the *Fermi* satellite and other experiments.

Hawking's ideas are now widely accepted, but they shocked many in the 1970s. Einstein's theory had seemed to imply that anything behind the event horizon would be locked in forever. Instead, Hawking showed that a black hole can evaporate, as if it were a hot liquid, and eventually disappear in a final explosion.

The information crisis

This led to a new puzzle, also raised by Hawking, in a landmark paper published in 1976. Think of a star collapsing to form a black hole (Figure 82). The star contains a huge number of atoms in complicated states. All of this 'information' is eventually hidden behind the event horizon. The black hole then starts to evaporate, emitting thermal radiation. Once the hole has completely evaporated and disappeared, all that is left is thermal radiation. Now here is the remarkable thing. The radiation that is left behind contains no detailed record of what fell behind the horizon as the star collapsed—apart from the total mass, rotation (angular momentum), and charge (if any). No further details remain of the collapsing star, apart from these three parameters. Many physicists find this paradoxical. The reader might object that this is no different from burning a book and finding that the ashes, smoke, and light produced contain no record of what was written in it. But actually, in principle, if we could follow the detailed microscopic structure of the ashes, smoke, and light produced there is no reason of principle why we could not reconstruct the original book. In contrast, in the case of an evaporating black hole, it seems that in principle, no matter how precise our measurements, the infalling information is lost forever.

This puzzle came to be known as the 'information loss paradox'. Attempts to solve it have spawned literally thousands of scientific papers.

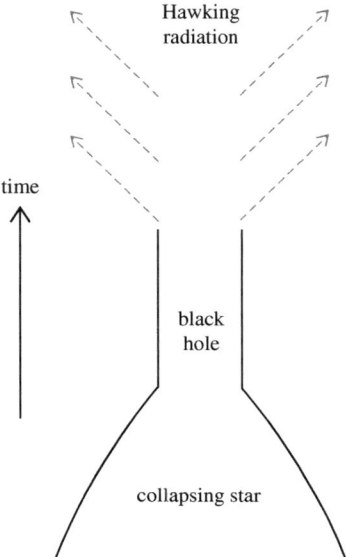

Figure 82. The formation and evaporation of a black hole. The complex material inside a collapsing star is converted entirely into thermal radiation, with no record remaining of the collapsed material apart from its mass, angular momentum, and charge.

Some physicists claim to have a solution, others say no. Whether the puzzle has been solved is controversial. Indeed, whether the puzzle really is a serious problem in the first place is itself controversial.

Let us look at the problem in another way. Imagine that the collapsing material is a quantum system with a specific quantum wave. Remember that the quantum wave changes with time according to the Schrödinger equation.[69] Like a water wave, the quantum wave moves around and spreads out. If we know what the wave looks like later, in principle we can calculate backwards in time and reconstruct what the wave looked like earlier. Now consider that collapsing material with a definite quantum wave. A black hole forms, which then evaporates, leaving behind nothing but thermal radiation. According to our best understanding, that radiation is described not by one specific quantum wave but by a mixture of different quantum waves. If we look at the final radiation, we are unable to reconstruct the

[69] We are considering a quantum system on a classical background space, so the usual Schrödinger equation applies (Chapter 8).

original quantum wave of the collapsing material. Put like this, Hawking's puzzle seems to violate the basic laws of quantum mechanics. According to those laws, one initial quantum wave evolves into one final quantum wave. And yet here the initial quantum wave has somehow evolved into a *mixture* of final quantum waves.[70]

It would be difficult to overstate the influence this paradox has had on the development of theoretical physics in the past half century. Attempts to solve it have inspired and motivated a huge amount of work in 'string theory'—a particular approach to theoretical high-energy physics that has dominated the field in recent decades. According to string theory, instead of being made of point-like particles moving in three-dimensional space, matter is really made of extended string-like objects vibrating in nine spatial dimensions (or a ten-dimensional spacetime). Space looks three-dimensional to us because the other dimensions are curled up ('compactified') into tiny circles too small for us to see. Alternatively, visible matter is confined to a three-dimensional 'brane' within the higher-dimensional space. String theory has some intriguing mathematical properties, even if its scientific status remains controversial. One of its more interesting applications has been to try to resolve the information loss paradox.

In string theory a mathematical relationship known as the 'AdS/CFT correspondence' suggests that the physics of black holes in a curved higher-dimensional space is equivalent to the physics of certain fields in a lower-dimensional flat space. But flat space means there is no gravity (remember that gravity curves space). So the higher-dimensional physics of black holes is somehow equivalent to a lower-dimensional physics with no gravity at all. And if there is no gravity, so the reasoning goes, there can be no gravitational collapse and no information loss. In other words, the perplexing physics of black holes appears to be equivalent to a simpler physics without gravity, in which information loss is impossible—suggesting that the information loss paradox is an illusion. The information trapped inside the black hole must somehow escape to the outside after all. But how? To try to answer this we can apply the AdS/CFT correspondence to translate the gravitational physics of black holes into the physics of certain fields without gravity, calculate what happens there, and then translate

---

[70] This is called a 'pure-to-mixed transition', which violates the basic quantum law of 'unitary evolution'.

back again. This is, however, much easier said than done. Despite strenuous efforts over the past twenty-five years, with many intriguing results, gaps remain. And the physical interpretation seems obscure. In particular, it is still not clear exactly how the information inside can be transferred to the outside. There are hints that, for the whole thing to work, the physics on the gravity side has to be nonlocal, potentially allowing superluminal signalling, which most physicists regard as impossible.[71]

In contrast, some other workers believe that information loss is instead a genuine physical phenomenon and that in this sense quantum mechanics breaks down when black holes evaporate. That was, in fact, Hawking's original viewpoint—though in his later years he appears to have changed his mind. As we have said, the whole subject is mired in confusion and controversy.

A new approach to information loss

This is genuinely difficult terrain. At this point, the reader will not be surprised to learn that pilot-wave theory offers a new approach to the information loss paradox.

We saw in Chapter 8 that fields in the vicinity of a black hole may be pushed away from quantum death by novel effects involving quantum gravity—effects which, however, make sense only in pilot-wave theory. As a result the outgoing Hawking radiation could break Born's formula. Let us pause for a moment and think about that. Normally, for particles with a given quantum wave, their statistical properties are fixed by Born's formula. For this reason there is only so much information they can carry. In other words: in the state of quantum death all the emerging particles carry the same quantum noise, and so there is only so much they can record about what fell into the black hole in the first place. But if the outgoing particles break Born's formula, their statistical properties can be more varied and more complicated—and so they can carry more information than usual.

And there is more. In Hawking's original calculations, fields in the space around the black hole are divided into two kinds of components, those that are moving inwards and those that are moving outwards. These 'ingoing'

---

[71] The idea that the physics of black holes must be in some sense nonlocal has been promoted in particular by the American physicist Steven Giddings.

and 'outgoing' fields turn out to be entangled. So now we have something interesting. If those entangled fields are pushed away from quantum death by the novel effects described in Chapter 8, they will provide a communication channel from the interior to the exterior. Remember from Chapter 5 that entangled nonequilibrium particles can be deployed for nonlocal or superluminal signalling. The same reasoning applies to entangled fields. And so information about what is happening inside the black hole can leak to the outside, casting doubt on the whole scenario of information loss. For example, if the outgoing nonequilibrium radiation hits a two-slit barrier, the anomalous pattern at the backstop could contain information about what fell inside the black hole (Figure 83).

And so we have a radically new mechanism whereby the interior of the black hole can communicate with the exterior. Because of nonequilibrium and the entangled fields, there is no genuine event horizon. It must be said, though, that this work is still in its infancy. More calculations are needed to work out exactly how much information can leak from the inside to the outside during the lifetime of a black hole, and to find out whether this will be enough to resolve the paradox of information loss. Still, it is fair to say that, as well as enabling our way out of quantum death, the pilot-wave theory of quantum gravity can also potentially illuminate one of the deepest problems of modern theoretical high-energy physics.

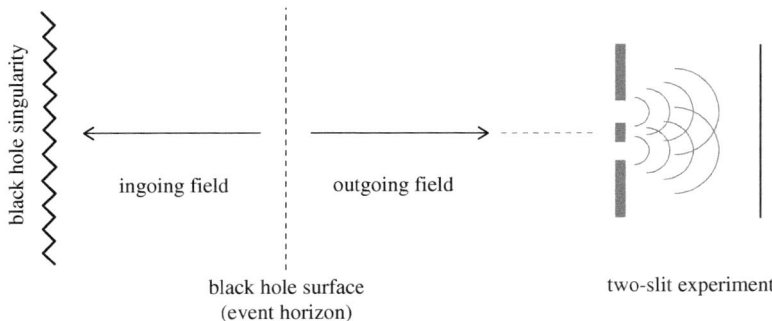

Figure 83. A new approach to information loss. When Born's formula is broken, the entangled ingoing and outgoing fields allow information to leak (superluminally) from inside a black hole to the outside. Hawking radiation could then contain information about what fell inside—information which might be visible as an anomalous pattern in a two-slit experiment.

Testing quantum physics with black holes

There is no doubt that black holes offer unique opportunities to stress test our most fundamental physical laws. These bizarre systems chip away at the edge of known physics, challenging basic principles of both quantum physics and gravitational physics. We have seen that they also offer what may be our best chance of escaping from quantum death. There is more to say about these fascinating objects—about the possible role they may play in the overall history of our universe, and about how they might be applied to push the limits of quantum physics.

We know that entangled particles can communicate instantaneously (even if the effects are currently obscured by quantum noise). We also know that the event horizon—the surface shrouding the mysterious interior of a black hole—prevents us from seeing into the interior because no outgoing signal can travel faster than light (or so physicists believe). If we wanted to puncture the horizon and peer into the interior, we would need to beat this supposedly fundamental speed limit. So here is something we might want to try. Let us take a pair of entangled particles and let one of the particles fall into a black hole. While that happens, let us monitor the remaining particle—which we keep here with us safely on the outside—and see if anything strange happens to it. In particular, let us check to see if our particle continues to obey the usual laws of quantum mechanics such as Born's formula. This can be done by repeating the experiment with many similar entangled pairs.

There are at least two reasons why it would be worth performing this experiment—if it were possible in practice.

First, even leaving aside the arguments made in this book, it is always a good idea to stress test our theories under genuinely novel conditions. The quantum physics of entangled particles has been tested in the lab over tens of metres (Chapter 3), and across Earth's surface over more than 1,000 kilometres (Chapter 10). In all of these experiments, there is no event horizon separating the particles. If the surface of a black hole really is a fundamental barrier, perhaps a particle on this side will show deviant behavior when entangled with a particle on the other side.

Second, recalling the arguments of Chapter 8, there are reasons to believe that Born's formula will break down when quantum gravity becomes important. This could happen, for example, close to the centre of a black hole where Einstein's theory predicts an infinite density of collapsed material (at the singularity). So it is reasonable to think that Born's formula will fail in the region close to the singularity. Fair enough, but how could we ever know? Here is a way. We take our pair of entangled particles and let one fall in. As the infalling particle approaches the singularity, it picks up violations of Born's formula from the fields in that neighborhood. Because the infalling particle is still entangled with our particle here on the outside, the two particles can communicate instantaneously.[72] As a result, our particle here on the outside also picks up violations of Born's formula. In other words: the entangled state straddling the event horizon can channel the anomalous noise from the inside to the outside[73]—resulting in a breakdown of quantum mechanics for the external particle, for example in a two-slit experiment (Figure 84). The external particle is knocked out of equilibrium by means of its entanglement with the infalling particle.

At this point the reader might wonder if we have lost touch with practical reality. How could we possibly perform such an experiment—in the absence of a black hole miraculously turning up in the lab? Well, it turns out that nature may have already provided us with a way to do it.

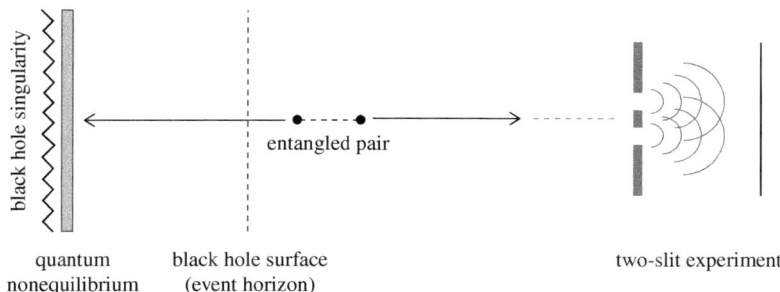

Figure 84. Probing a black hole with entangled particles. We drop one particle into the hole and closely monitor the other. Violations of Born's formula on the inside can be transferred superluminally to the outside, where they may be revealed as anomalies in a two-slit experiment.

---

[72] Technically, again, 'instantaneously' is defined with respect to a time parameter associated with a preferred slicing or foliation of spacetime.

[73] Calculations illustrating this effect have been done in collaboration with Adithya Kandhadai.

A quantum experiment with a black hole

Let us look a bit more closely at what is known about black holes. We have said that a black hole can form when a star collapses. For that to happen, in theory the star has to begin with at least about three times the mass of the Sun—in practice probably rather more, since towards the end of its life a star can blow off much of its mass when it explodes in a supernova (what is left behind can then collapse to a black hole). In any case, there are plenty of stars sufficiently massive that they are likely to end their lives as black holes.[74] It is estimated that our galaxy contains millions of such 'stellar' black holes, ranging from about ten to about thirty times the mass of the Sun (precise estimates vary), and spread around the galaxy much as visible stars are. These dark objects are about 30 to 90 kilometres across (as measured by the radius of their event horizon beyond which light cannot escape). In the past two or three decades attention has shifted to a different kind of black hole—huge objects of at least a *million* times the mass of the Sun. These are known as 'supermassive' black holes. One of these monsters lurks at the centre of our galaxy (coinciding with the object known to astronomers as Sagittarius A*). It is believed that almost every large galaxy has a supermassive black hole at its centre, which probably grew from the merging of smaller black holes or by accreting matter from the surroundings.

Now it is interesting to ask how these objects were discovered. After all, by definition a black hole does not emit light and so cannot be seen directly. Even so we can tell a black hole is there by its gravitational effects. At the simplest level, we might notice that certain stars have peculiar high-velocity orbits, which can only be explained if they are being sharply tugged by some small and extremely massive object in their neighborhood. In fact this was important evidence for the supermassive black hole at our galactic centre, which earned German astrophysicist Reinhard Genzel and American astrophysicist Andrea Ghez the 2020 Nobel Prize in Physics (shared with Sir Roger Penrose, whose theoretical work in the 1960s proved that black holes are an inevitable consequence of Einstein's theory of gravity). But tugging sharply on nearby stars does not tell us much in the way of detail about the black hole itself. For that we need to

---

[74] We have also seen that much smaller 'primordial' black holes can form in the early universe, but those are not our focus here.

probe the highly curved space close to the event horizon. Short of sending a spacecraft all that way (it is more than 3,000 light years from Earth to the nearest known black hole), how can we possibly do that? Remarkably, nature herself has provided us with a way. To explain it, we will need a bit of high-energy astrophysics.

Consider what happens in the space around a black hole. Any matter in its vicinity is likely to be drawn towards it, eventually falling inwards. In practice, it often happens that interstellar material (including gas and dust) is drawn towards the hole but instead of falling straight in the material swirls around and around and around, before eventually spiralling inwards—like water draining into the hole of a bathtub. This swirling material takes the shape of a thin flat disc known as an 'accretion disc' (Figure 85). As the material swirls inwards it heats up. Some of it becomes so hot that it emits X-rays—lots of X-rays, with a range of frequencies. These can be detected by appropriate X-ray telescopes housed on satellites high above the obscuring atmosphere of Earth. Since the 1990s, in fact, X-ray astronomy satellites have been studying black hole accretion discs by monitoring the frequency spectrum of the emitted X-rays.

Now we need to drill down into a bit of detail. The accretion disc contains a variety of material, including atoms of iron. Above the surface of the disc, in the region close to the black hole, there is an extremely hot 'corona' producing X-rays with a broad range of frequencies. Some of those rays hit the surface, exciting the iron atoms, which then re-radiate X-rays with a definite frequency—what physicists call a sharp spectral line. And this is why, when satellites measure the X-ray emission from an accretion disc, they find a smooth spread of frequencies punctuated by a sharp line emitted by iron atoms. We might imagine the hiss of white noise, with a smooth spread of frequencies, punctuated by a single sharp tone at one frequency.

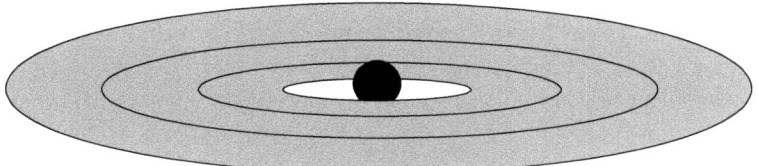

Figure 85. A black hole with an accretion disc.

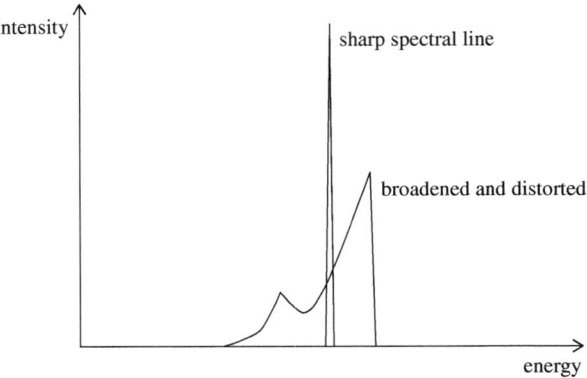

Figure 86. A natural (or 'intrinsic') sharp spectral line compared with an example of a broadened and distorted line.

How does this help astrophysicists probe the curved space close to a black hole? Here is how it works. We have said that iron atoms in the accretion disc emit X-rays with a precise frequency. But if the emission occurs close to the black hole, to reach us the X-rays have to climb out of the gravitational field around the hole. In the process, the X-rays lose energy and their frequency goes down.[75] Exactly how much energy is lost will depend on how close the atom was to the surface of the black hole. So when we measure the X-rays coming from different atoms, at different places in the accretion disc, they will have slightly different frequencies. The actual spectral line we see is then not quite sharp but smeared or 'broadened'. Similarly, the speeds at which the atoms are moving also affect the emitted frequencies, and those speeds depend on the details of the gravitational field near the hole. In short, the actual 'iron line' our satellite detects is somewhat broadened and distorted. And here is the punchline: the precise *shape* of the line provides us with a telltale signature of the curvature of space close to the surface of the black hole (Figure 86).

These details were first worked out in 1989 by British astrophysicist Andrew Fabian and collaborators. A few years later, in 1995, a broadened and distorted iron line was first observed, by the Japanese *Advanced*

---

[75] This can also be understood in terms of 'gravitational time dilation'—the effective slowing down of time in a strong gravitational field.

Satellite for Cosmology and Astrophysics, in X-rays from the centre of the 'active galaxy' known as MCG–6-30-15 (about 100 million light years from Earth). From the shape of the line, it has since been possible to deduce that the central supermassive black hole is spinning rapidly. Similar results have been found for some stellar black holes in our galaxy.

After that astrophysical detour, we can return to our quantum experiment. Remember we want to take a pair of entangled particles, drop one inside a black hole, and see what happens to the other (Figure 84). But the closest known black hole is more than 3,000 light years away. So how can we possibly perform our experiment? There may be a way.

We have seen how astrophysicists are able to detect X-ray photons emitted by iron atoms close to the surface of some black holes. Those photons are emitted one at a time. Now imagine what might happen if an atom close to the surface of a black hole happened to emit *two* photons at a time. In the right circumstances the photons will be entangled. And they will be emitted in different directions. One of the photons could fall into the hole, while the other flies off in the other direction, away from the hole, and is eventually detected by our X-ray satellite (Figure 87). We would then have something like the experiment we are looking for. When the outgoing photons reach our satellite, they could be tested to see if they obey Born's formula. This could be done by letting those photons hit a two-slit experiment: perhaps the usual wavy pattern will be blurred or degraded in some way. Alternatively, we could test the photons' polarisation probabilities. Either way, if the detected photons are found to break Born's formula, it is game over for quantum mechanics.

So now it all hinges on us being able to find atoms close to a black hole that emit not single photons but entangled pairs. This, again, is not as fanciful as it might sound. In a typical one-step atomic decay, an atom makes a transition from one energy state to a lower energy state, emitting a single photon in the process. But two-step atomic decays are also quite common: starting at some high energy level, the atom drops to a lower level while emitting one photon, and then the atom quickly drops to an even lower level while emitting yet another photon. In such a two-step decay, an atom can emit two photons at essentially the same time. And—here is the interesting point—the pair of photons can be entangled. In fact, this is how entangled photons are commonly produced in the lab. We excite an atom and then let it decay in a two-step process (known as a 'cascade emission') so that it emits a pair of entangled photons. And so, to perform our experiment, we

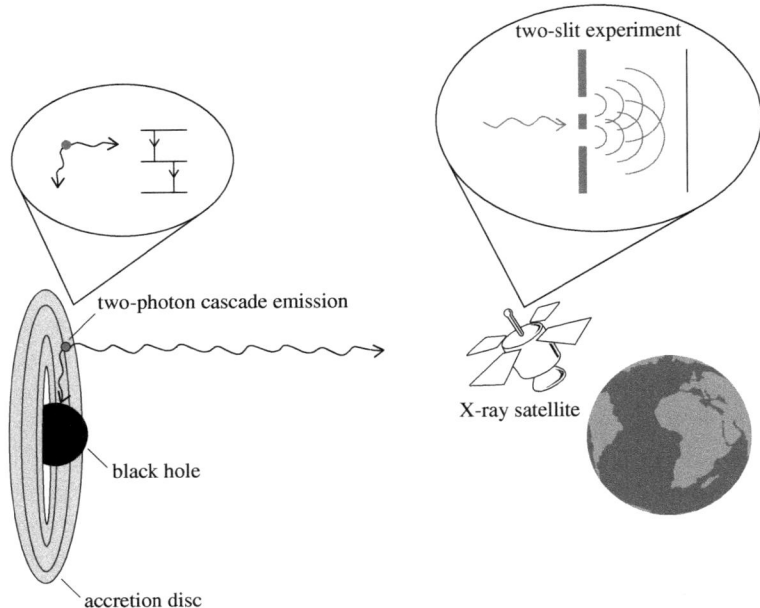

Figure 87. A quantum experiment with a black hole. An atom close to the hole emits two entangled photons. One falls in while the other reaches a distant X-ray satellite—where it is tested to see if it obeys Born's probability formula (for many similar photons).

just have to find a naturally occurring two-step (or cascade) emission in an accretion disc, close to the surface of a black hole.

Can this really be done? Let us look at those iron atoms in a bit more detail. We have said, in simple terms, that the iron atoms get excited when hit by X-rays from the hot corona above the disc. More precisely, what really happens is that an electron is knocked out of the innermost region of an iron atom, creating a 'vacancy' into which a higher-level electron can drop, losing energy in the process and emitting an X-ray photon (which can then be seen by our X-ray satellite). But this leaves another vacancy higher up in the atom, into which a further outer electron can drop, emitting another photon. This is in effect a two-step decay or cascade emission. Basic atomic physics indicates that only a small fraction (less than 1%) of the iron line photons we see can be accompanied by cascade partners. So the anomalies we are looking for can exist only for a small fraction of the photons we receive. Clearly, if there are anomalies they will be diluted and difficult to measure. Even so, in a sense our experiment is already out there, set up for us by natural atomic and astrophysical phenomena.

For the experiment to be worthwhile, we must be sure that the photons are being emitted by atoms close to the surface of the black hole—so that one of the photons is likely to be captured (falling behind the horizon) while the other flies outwards eventually reaching our satellite. How can we be sure that the emission is taking place close to the surface? Happily, this has an easy answer. Remember the photons lose energy as they climb out of the gravitational field of the hole. The closer to the surface they are emitted, the harder the climb and the more energy is lost. And the more energy is lost, the lower will be the frequency of the escaping photon. The message is clear. If we are particularly interested in photons that were emitted as close as possible to the surface of the black hole, we need to focus on photons at the low-frequency or low-energy end of the observed broadened iron line (Figure 86). Equivalently, we need to look at the long-wavelength end of the line—known as the 'extreme red wing'. For certain black holes the extreme red wing of the iron line is known to originate from just outside the horizon (at a distance less than twice the radius of the black hole). Some of those photons will have entangled partners that were captured by the black hole.

It seems fair to say that the essential ideas are in place: the experiment could be conducted with incoming photons from the extreme red wing of an iron line emitted from the accretion disc of a black hole. There are some practical issues to consider though. How strongly entangled are the photon pairs likely to be? This has yet to be studied—it could be tested experimentally in the lab (by firing X-rays at iron atoms and studying the cascade emissions). We might also ask which black holes would be good candidates for the experiment. No significant iron line has been found for the much-studied supermassive black hole at the centre of our galaxy, whose X-ray emission seems unusually faint. But a clear iron line has been seen, as noted, for the supermassive black hole at the centre of the distant galaxy MCG–6-30-15. Should we wish to stay within our galactic neighbourhood, a clear iron line has been seen for the black hole known as GX 339-4, located in our galaxy at a distance of about 25,000 light years from Earth. Finally, even if some of the outgoing photons really do violate Born's formula, they might undergo quantum relaxation during transit towards Earth, owing to small interactions with the intergalactic and/or interstellar media, in which case we will not see anything unusual by the time the photons reach us.

Clearly, even if our ideas are essentially correct, the success of this experiment will be a long shot. Still, it would be worthwhile as an extreme stress test of quantum mechanics, under the most unusual conditions studied so far.

Reprocessing the quantum universe

Over the past half century many physicists have pinned their hopes on black holes as harbingers of the next revolution in physics. In the early 1970s the visionary American theorist John Wheeler—who coined the term 'black hole'—called gravitational collapse 'the greatest crisis in physics of all time'. Einstein's theory of gravity predicts that matter can collapse to a point of infinite density—the singularity inside a black hole—a point where the equations of physics no longer make sense. As Wheeler put it, it seems that physics has predicted its own demise, and yet physics is by definition that subject which must go on no matter what. Wheeler was particularly perturbed by the idea that the whole universe might eventually reach a point of maximum expansion, recontract, and undergo a universe-wide gravitational collapse to a singularity (which can happen if the universe contains enough mass). He suggested that in a sense the universe might be renewing—or 'reprocessing'—itself by means of gravitational collapse. In what at the time must have seemed a fit of idle speculation, Wheeler suggested that the whole universe might undergo cycles of expansion and recontraction, where after each collapse the universe emerges fundamentally changed, containing new kinds of particles and different values of the 'fundamental physical constants' (in particular the numbers that measure the strengths of the fundamental forces).

Some of Wheeler's concerns seem out of date now. Current evidence suggests that our universe does not contain enough mass to cause it to recollapse. Instead it is expected to expand forever. And inflationary cosmology has many new things to say about the overall history of our universe (which may be just one of an infinity of inflating universes). In any case, these days it is generally believed that singularities will be avoided by new effects from quantum gravity or string theory. Even so, there is something haunting about Wheeler's vision. In his 1997 book *The Life of the Cosmos*,

Smolin suggested that black holes could produce 'baby universes' (on the 'other side' of the interior) with slightly different values of the fundamental physical constants. He argued that an analogue of Darwinian natural selection would make our universe fine-tuned to produce as many black holes as possible. On Smolin's theory, black holes not only reprocess the basic physical constants, they also give birth to new universes and provide an opportunity for something like natural selection to operate on a cosmological scale.

In this book we have found that, if our ideas are correct, in a certain sense black holes 'reprocess' the quantum universe—they take in systems in a state of quantum death and return systems in the pre-quantum state where quantum theory fails. It is instructive to consider this possibility in a broader context.

According to quantum gravity, at least as understood in this book, our universe necessarily began in a pre-quantum or nonequilibrium state where quantum mechanics was invalid. As we saw in Chapter 8, this must be so because the quantum wave defined by quantum gravity cannot represent a bona fide probability. And so Born's formula is meaningless in the deep quantum-gravity regime from which our universe presumably emerged. Soon afterwards, however, as the universe expands and cools, conventional quantum waves emerge. Then begins the process of quantum relaxation—whereby the universe moves towards its present equilibrium state of quantum death, with the ubiquitous quantum noise we see today. For much of this book we feared that we might be trapped in this state forever. But then we understood that, as black holes form and eventually evaporate, they seem able to 'revive' the universe and create particles that are again in a pre-quantum or nonequilibrium state where quantum mechanics fails. In a word, gravity can save us from quantum death.

We might put it like this. At the beginning of time, gravity ensures that the universe starts out in a state of quantum life beyond Born's formula. The universe then 'dies' by quantum relaxation, leaving it in a state of quantum death from which there is seemingly no escape. Later on, gravity again saves the day: matter undergoes the dramatic process of gravitational collapse to form black holes, which slowly evaporate and eventually explode in a final burst of radiation that marks 'quantum rebirth' (Figure 88). The quantum universe is reprocessed by gravitational collapse, reviving the universe from quantum death and transcending quantum

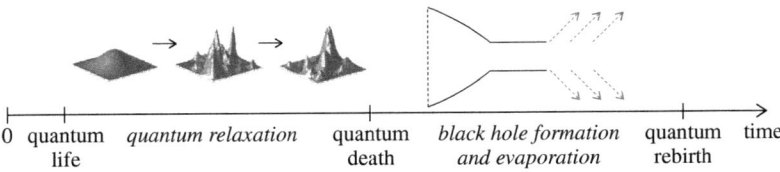

Figure 88. Reprocessing the quantum universe. From quantum life to quantum death, and back again, mediated by the effects of gravity.

physics altogether—in a sense vindicating something of Wheeler's original vision.

Similarly, if our universe was indeed born from inside a black hole (as in Smolin's scenario), and if quantum gravity was involved in this process, then we can again expect our universe to have started in a state of quantum life, with relaxation to quantum death taking place only later. Then, as black holes form in our universe, any new universes they may create will again start in a state of quantum life, and the process will be repeated ad infinitum. What the implications of this might be, for quantum cosmology and beyond, is left for future research.

10
Beyond Quantum Technology

The quantum space race

In 1957 the Soviet Union launched *Sputnik 1*—the world's first artificial Earth satellite. The steady beep of its radio pulses could be heard all over the world, provoking both fear and excitement. The sense that the Russians had surpassed the Americans triggered what came to be known as the 'space race'—in which the United States and the Soviet Union vied for supremacy in the new frontier of space. Today we are in the midst of a much more complicated 'quantum space race', with many actors and competitors, including universities, government intelligence agencies, corporations, industry, and the military.

In 2016 China launched *Micius*, the world's first quantum communications satellite. Observers on the ground linked to the satellite can employ 'quantum cryptography' to send each other messages. The speed of these messages is limited by the usual speed of light. But there is a novelty: the messages are unhackable by any spy or eavesdropper. In 2019 Google announced that it had achieved 'quantum supremacy'. This meant they had built and operated a quantum computer that was far faster than any ordinary or classical computer (at least for a specific task). The age of quantum computing had arrived, and classical computers could never catch up. While this claim was quickly disputed, by 2024 it appeared that quantum supremacy had indeed been achieved. These are recent milestones of what is often called—somewhat misleadingly—the 'second quantum revolution'. It is a revolution of sorts, but not of physics. This new technology is based on the usual laws of textbook quantum mechanics. In the 1920s we had a quantum revolution in science. Now we are witnessing a quantum revolution in technology.

The rise of quantum technology began in earnest in the 1990s, with the widespread realisation that some of the seemingly philosophical peculiarities of quantum physics had practical applications. Quantum

Beyond the Quantum. Antony Valentini, Oxford University Press. © Antony Valentini (2025).
DOI: 10.1093/oso/9780198853749.003.0011

uncertainty and entanglement can be applied to encrypt and decrypt messages in total secrecy. They can also be harnessed for superfast computation. Thirty years later we find ourselves in a world-wide race for quantum technology, powered by billions of dollars of research funding.

One major goal is to build a general purpose quantum computer. Such a machine is probably at least some years away, perhaps even decades. While opinion is still divided as to how generally useful it will turn out to be, it will certainly excel at some specialised tasks. Important potential applications include breaking widely used methods of encryption, the design of new drugs and materials (by rapid computer simulations), the efficient solution of complex logistical problems, and high-speed financial services. Some of these applications may arrive in the near future, others in the longer term. On one estimate, quantum computer technology is likely to have a global market value of $1 trillion by 2035.

But practical quantum cryptography is already here—and is being mobilised to build a supposedly unhackable 'quantum internet'. One year after its launch, *Micius* successfully shared entangled pairs of photons between ground stations separated by more than 1,200 kilometres (Figure 89). As we have discussed at length, in quantum mechanics entangled particles cannot be employed for superluminal signalling. Even so, according to that same quantum mechanics, this set-up can provide a global communication system that is impenetrable to hackers or eavesdroppers.

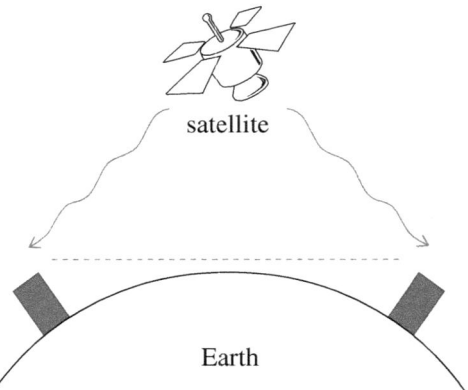

Figure 89. The quantum satellite *Micius* shares entangled photons between remote ground stations on Earth, enabling the transmission of unhackable messages between them.

As we will explain, the idea is to apply the peculiar properties of entanglement to share 'secret keys' that can be used to encrypt and decrypt messages—a feat accomplished with *Micius* in 2020. The system is advertised to be impenetrable not merely in the sense of being very difficult to hack but in the sense of being *impossible* to hack. To hack or eavesdrop on the system would violate the laws of quantum mechanics. In other words: to steal these quantum secrets, we need to violate the known laws of physics.

After *Micius*, similar quantum satellites are being planned or launched by other nations, including a joint venture between the UK and Singapore. However, while orbiting satellites are likely to play a crucial role in the imminent quantum internet, terrestrial fibre-optic cables (transmitting photons between remote locations) will also be important. The cities of Jinan and Qingdao in China are already connected by such a 'secure quantum data link'—where photons sent through a 511-kilometre fibre-optic cable carry a secret key that can be used to encrypt and decrypt messages in ultimate unhackable secrecy. Similar links are being developed across the world. If one day quantum computers become fully operational, they will be included in the quantum network, with potentially dramatic consequences.

Like most new technology, a completely unhackable communication system can be harnessed for both good and ill. It will doubtless be a boon to private citizens who wish to avoid being snooped on by oppressive governments—as well as to criminals who wish to evade the law. There is also a military dimension. Today conflicts are increasingly likely to be fought in cyberspace. In the near future the domination of the world will surely include the domination of information—control of its content, transmission, and security. Significantly, the AUKUS defense and security pact between Australia, the UK, and the US, signed in 2021, includes cooperation on quantum technology.

Betting on a theory that makes no sense

Quantum mechanics promises communication in total secrecy. In all the excitement, however, a basic fact tends to be overlooked. Quantum mechanics does not make sense. It will inevitably be replaced by a deeper theory that does make sense. What happens then to the security of the quantum internet?

In effect governments, corporations, industry, and the military are placing bets of billions of dollars that the laws of quantum mechanics are rock-solid reliable forever—even though those laws make no sense, and even though in the past our physical laws have always turned out to be limited and to fail at some point. If the bet is lost and the laws of quantum mechanics eventually fail, the supposedly unassailable security of the quantum internet could be breached, in which case the whole system will come crashing down.

People have always tended to believe that the physical laws and principles that they are familiar with in their lifetimes are final and unassailable. It is only human to think, or perhaps hope, that our present beliefs and habits are somehow permanent. But here there is something more. If Newton's laws fail or become inaccurate in some regime (for example, at the atomic scale) they still work well in their own domain: we can still apply those laws to design rockets without fear of them crashing. In the case of the quantum internet, where the whole point is absolute security, if the laws fail even by only a little bit the entire system can crash. We saw in the Prologue that quantum mechanics is one of the most confused and nonsensical scientific theories in history. So why are so many scientists, engineers, corporations, governments, and funding agencies willing to bet billions of dollars on its ultimate validity?

The inconsistencies of quantum mechanics have been swept under the rug, while a coalition of physicists, computer scientists, and communications engineers develop its novel applications to information technology. The practical implications are potentially immense. There is an understandable impulse not to let seemingly philosophical quibbles get in the way when we are on the verge of important technological breakthroughs. But what if those philosophical quibbles threaten to undermine the entire enterprise?

There have also been attempts to repackage quantum mechanics to make it sound as if physics is fundamentally about 'information'—a word which, in effect, replaces the older word 'observation'. Here the age-old conflict between perception and reality rears its head once more. A physics of information amounts to a physics of observation or perception only, with no regard for any underlying reality. To claim that reality is fundamentally about information is as confused as the claim in the 1920s that reality is fundamentally about observation. As Bell pithily put it:

Information? *Whose* information? Information about *what*?

Despite some claims to the contrary, the development of quantum technology has done little to advance our understanding of quantum reality. In fact, it has arguably distracted attention from the central problem, which is to understand what is actually happening in the quantum world even when no experimenters are there to observe it—or to acquire information about it.

In this book we have argued that quantum mechanics is not a fundamental theory. It merely describes a special state of quantum death in which all things are subject to the same uncertainty noise—an all-pervading fog that prevents us from seeing and controlling the quantum world accurately. Behind the fog, there lies an unexplored reality in which all things are deeply and instantaneously connected. In principle there is a much wider physics—'beyond the quantum'—in which the usual fog is avoided. In this new physics, the uncertainty principle can be beaten and entangled particles can be employed for practical superluminal signalling. We have seen that this new physics may have existed in the early universe. And it may still exist today, in relic particles from early times, or in radiation from evaporating black holes. Subquantum matter would break Born's quantum probability formula and could be harnessed to perform tasks that are now regarded as impossible. All this raises two questions. How might this new physics affect the development of quantum technology? And what even more radical technology might be made possible by the discovery of subquantum matter?

Hacking the quantum internet

Let us begin by looking at how quantum secret messaging is supposed to work, why it is believed to be absolutely unhackable, and how the system can in fact be hacked if we manage to escape from quantum death.

First, we should clarify a common misunderstanding. In these techniques quantum effects are not used to send an actual message. Instead quantum effects are used to create and share a secret key. This can be thought of as something like a password, usually written as a string of 1's and 0's, for example 110100. The key is supposed to be known only to the sender and receiver, who in this context are traditionally called 'Alice' and 'Bob'. As shown in Figure 90, Alice applies the key to encrypt

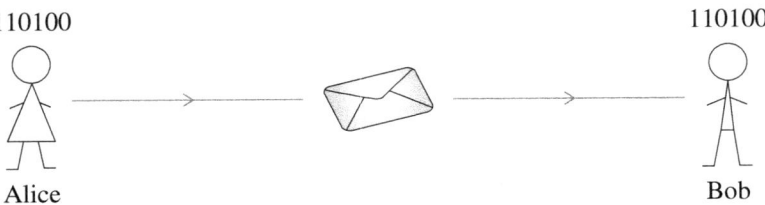

110100   110100

Alice   Bob

Figure 90. Secret-key cryptography. Alice applies a key (say 110100) to encrypt a message, which she then sends to Bob by conventional means. Bob applies the same key to decrypt the message. This can work only if Alice and Bob can secretly share their key in advance.

(or scramble) the message, which she then sends by conventional means. Upon receipt Bob applies the same key to decrypt (or unscramble) the message, which he is then able to read. For this to work in secret, Alice and Bob need to (somehow) share their key in advance—without anyone else finding out what the key is. So what this is really about is 'quantum key distribution': how to apply quantum effects to create and share a secret key.

There are different methods or 'protocols'. We begin with the one published in 1991 by the Polish-British physicist Artur Ekert.[76] The essential idea is quite simple. Alice and Bob share a pair of entangled particles—say electrons—which have been prepared with a special quantum wave (a singlet state). Each electron passes through a magnetic field and is deflected upwards or downwards (Figure 91). As we saw at the end of Chapter 3, this is often called a 'measurement of spin', but what matters here is that each particle is deflected up or down. The experiment is repeated many times with that same initial quantum wave. If the two magnetic fields are aligned, the particles are always deflected in *opposite* directions. If Alice's particle moves up, Bob's particle moves down; while if Alice's particle moves down, Bob's particle moves up. Whether each particle actually moves up or down is, according to quantum theory, purely a matter of chance. According to Born's formula, applied to this particular quantum wave, if the experiment is performed many times half of Alice's particles will move up and half will move down (with Bob's particles moving in the opposite direction in each case). So now here is the trick. Let us agree in advance that if Alice's particle comes out up or down she writes a 1 or 0, respectively, while (in the opposite sense) if Bob's particle comes out down

---

[76] The *Micius* satellite employs a more practical version of Ekert's original protocol.

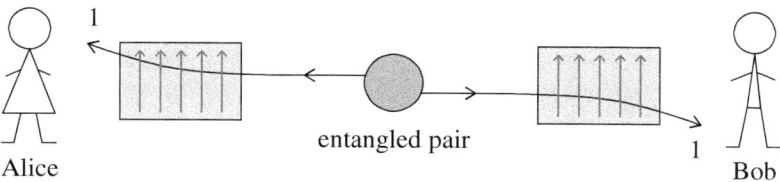

Figure 91. Ekert's 1991 protocol. Two entangled particles are each deflected by a magnetic field. In appropriate conditions the particles are always deflected in opposite directions, thereby generating the same key on each side.

or up he writes a 1 or 0, respectively. What will happen? If the experiment is repeated, say six times, Alice will write a (supposedly random) six-digit string such as 110100, and Bob will write exactly the *same* string 110100. In other words, the same key has been created for both parties. And of course the key can be made as long as we like by repeating the experiment enough times.

We might think of it like this. Imagine that Alice and Bob each toss a fair coin with an even chance of getting heads or tails. They repeat this many times. By some miracle, whenever Alice gets heads Bob gets tails, and whenever Alice gets tails Bob gets heads. If Alice writes a 1 when she gets heads and a 0 when she gets tails, and if Bob does the opposite, they will both end up writing exactly the same string of 1's and 0's—which they can then use as a shared secret key for cryptography.

As the reader will have noticed, to say that the key has been 'shared' or 'distributed' is a bit misleading. It would be more accurate to say that the same key has been *created* at two different locations. Note also that, because of quantum entanglement, Alice and Bob can be as far away from each other as they like and it will still work just the same.

But what if an eavesdropper—traditionally called 'Eve'—measures the spins (the up or down trajectories) *before* the particles reach Alice and Bob? Will she then know in advance what the key is going to be? Yes, indeed she will, and this would destroy any secrecy. And that is why there is more to Ekert's protocol than we have said so far.

To guard against eavesdropping, we can add the following trick. Instead of making sure their magnetic fields are always aligned for each pair of particles, Alice and Bob can orient their magnetic fields in different ways so that sometimes they are aligned and sometimes not. The cases where the fields happen to be aligned are used to generate the shared key as we have described. The cases where the fields happen to be *not* aligned are used to test for the presence of an eavesdropper. Here is how. When the fields are not aligned the particles no longer come out moving in opposite directions, but even so there is a statistical correlation between the results.[77] That correlation is determined by the quantum wave. Now—here comes the point—if Eve tries to eavesdrop on the particles (by measuring them in some way) that will change the quantum wave and upset the correlation. In Ekert's protocol, if the non-aligned cases are found to be not correlated as they should be, the alarm bell rings and we know that Eve is up to something. If instead we see the right correlations for the non-aligned cases, we know we are safe, and we can proceed with the aligned cases to generate our shared secret key.

That is all there is to it. If these precautions are taken, Alice and Bob can use their shared key to message each other in total secrecy, confident that they and they alone are in possession of the key. The system is totally and absolutely unhackable.

Or is it? Remember that the 1's and 0's in the key are a record of whether each particle came out up or down after it passed through a magnetic field. If Eve somehow knew in advance which particle will come out up and which particle will come out down, she would already know what the key is going to be. To avoid setting off the alarm, she would have to accomplish this without disturbing the quantum wave. According to the textbooks this is impossible. The outcomes—up or down—are not determined by anything. They are the result of pure chance. And any attempt to gain more information will inevitably disturb the quantum wave. So it really does look totally unhackable—if we believe quantum theory.

If instead we believe pilot-wave theory, the outcomes are not the result of pure chance but are determined by de Broglie's law of motion. Specifically, if Eve knew the exact initial positions of the particles inside the quantum wave, she could apply de Broglie's equation to calculate where the particles will go (Figure 92). And so she could predict what the key is going to be. The hard part is knowing the initial positions of the particles—remembering that this has to be done without disturbing the quantum wave. We have already seen how to do this in Chapter 5, when we discussed subquantum measurement. *If* Eve has equipment constructed from nonequilibrium or subquantum matter, with negligible statistical noise, she

---

[77] See our discussion of Bell's theorem at the end of Chapter 3.

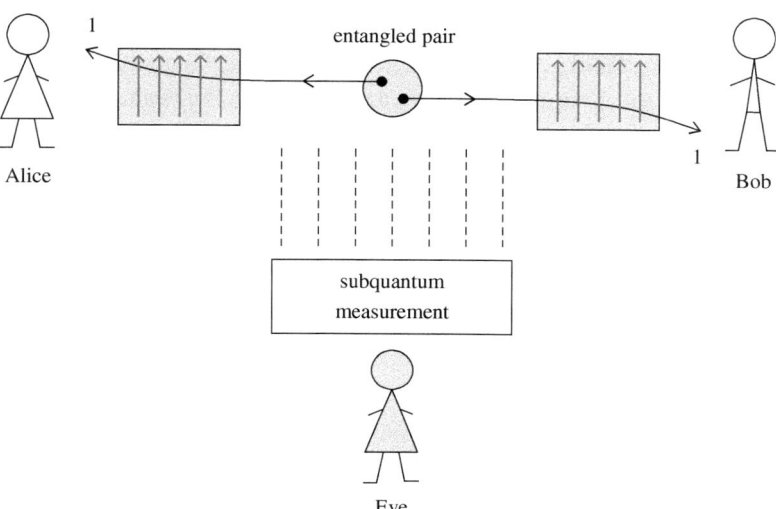

Figure 92. Hacking the quantum internet—or eavesdropping on quantum key distribution. With the aid of subquantum matter, Eve is able to circumvent the uncertainty principle and measure the initial particle positions without disturbing the quantum wave. She is then able to predict in advance where the particles will go and what the 'secret' key is going to be.

can use it to monitor the positions of the particles without disturbing their quantum wave. In that case she can hack into the system, simply by already knowing in advance how the particles are going to move and what the key is going to be (Figure 92).[78]

We might think of it this way. Imagine someone throwing a tennis ball. If we cannot see the ball as it leaves their hand, we will not know exactly where it was thrown from, in what direction, or at what speed—so obviously we will have no way of predicting where the ball will land. But if we *can* see the ball as it leaves their hand, we can apply Newton's laws to work out its motion and indeed predict where it will land. Normally, in the state of quantum death (which is the only state allowed by quantum mechanics), we are unable to watch the 'ball' (or electron) as it is thrown and so

[78] In practice, Ekert's protocol is implemented not with electron spins but with photon polarisations. In pilot-wave theory, photons are described not as particles but as energetic properties of moving fields. To hack the system, Eve would have to monitor the underlying moving fields. The details differ from the electron case but the principles are the same.

we cannot tell where it will land. But if we have equipment that is outside the state of quantum death—with less noise than usual—then we can watch the 'ball' (or electron) as it is thrown and (with a bit of calculation) tell in advance how it will move and where it will land.

So that is how Eve can hack the quantum internet (at least in Ekert's version). She needs a supply of subquantum matter. She can make use of it to monitor the entangled particles employed by Alice and Bob, and in this way she can predict in advance how the particles will move and what the shared 'secret' key is going to be—without disturbing the quantum wave and without setting off any alarm bells. The unhackable can be hacked after all.

It is interesting that in a sense Ekert's original paper anticipated this kind of attack, claiming however that it would be impossible. This is what Ekert had to say:

> The eavesdropper cannot elicit any information from the particles while in transit . . . simply because there is no information encoded there. The information 'comes into being' only after the legitimate users perform measurements . . ..

In other words: the particle trajectories (up or down) come into being only when Alice and Bob observe them. The act of measurement creates reality. This reminds us of how university physics students are taught to think about the simple experiment shown in Figure 44 (Chapter 3). A particle is fired at a barrier with only one hole. The incoming wave diffracts at the hole and spreads all over the backstop. There is no particle anywhere. And yet, when we take a look—*pop!*—the particle appears seemingly out of nowhere. The particle 'comes into being' by the power of the observer.

So now we can see how the security of the quantum internet depends on the widespread belief that there is no underlying quantum reality—that the observers' *perceptions* are in effect the only reality. Our most intimate secrets will soon be entrusted to this belief.

We have said that Ekert's is not the only protocol for quantum key distribution. Let us take a look at another one, published in 1992 by the American physicist Charles Bennett.[79] In this method there are no entangled particles.

---

[79] This is a simpler version of a 1984 protocol by Bennett and Canadian physicist Gilles Brassard. The original ideas go back to unpublished work in the late 1960s by the far-sighted American-Israeli physicist Stephen Wiesner.

Instead, Alice sends Bob a stream of particles, one at a time, where each particle has one of two possible quantum waves—let us call them 'wave A' and 'wave B'. The two possible waves are chosen so that, in ordinary quantum mechanics, they cannot be reliably distinguished.[80] Bob performs certain kinds of measurements on the particles he receives. The labels (A or B) on some of the waves are used to define the shared key. Without going into details, let us just say that the security of this method rests on the eavesdropper Eve being unable to distinguish wave A from wave B reliably for each particle. And if she does try to distinguish the waves she will inevitably disturb them—causing changes that could be detected by Alice and Bob.

We have not delved into the details of Bennett's protocol because we wish to focus on the most important point: that its security rests on Eve's inability to distinguish wave A from wave B reliably. Why should it be so difficult to tell the difference between two different quantum waves? To see what is going on, let us think about the vivid example of Schrödinger's cat. As we saw in the Prologue, according to quantum mechanics the cat could be dead, it could be alive, or it could be in a superposition of both dead and alive. Each of these states is represented by a particular quantum wave. Now imagine someone tells us that the cat has been prepared in one of two states, 'dead' or 'alive'. Can we tell which quantum wave it is? Well, this one is easy: we look at the cat, and if we find it dead then the wave was 'dead', whereas if we find it alive then the wave was 'alive'. Obviously. But now let us try something different. Let us say the two possible quantum waves are 'dead' and 'dead + alive'. Can we tell the difference? Well, not necessarily. If the wave is 'dead' we will definitely find the cat dead, while if the wave is 'dead + alive' there is a 50% chance we will find the cat dead and a 50% chance we will find the cat alive. So what happens? If we find the cat alive, then the wave must have been 'dead + alive'. Fair enough. But if we find the cat dead, then the wave could have been 'dead' but it could *also* have been 'dead + alive'—we have no way of telling which it was.

We have given the example of a cat but in everyday practice in the lab this sort of thing happens with particles, or atoms, which might have two energy states which we can call E_1 and E_2. Let us say our particle has quantum

---

[80] Technically, the quantum states are 'non-orthogonal'.

wave 'E_1' or '$E_1 + E_2$'. The same reasoning applies. If we measure the energy and we find E_1, we will not know which quantum wave it was.

So this peculiarity of superpositions can lead to a situation where we cannot reliably distinguish two different quantum waves—if they have a component in common. This is known in the jargon as the problem of reliably distinguishing 'non-orthogonal states' (for a single system). According to quantum mechanics, this task is actually *impossible*. If we are told that a single particle has been prepared in one of two possible non-orthogonal states, there is no way we can reliably tell which state (or quantum wave) it really is. And this fact underpins the security of the quantum internet in the version given by Bennett. If Alice and Bob send each other particles which have been randomly selected to be in one of two possible non-orthogonal states 'A' or 'B', it is impossible for Eve to know for certain which states have been sent and so she cannot find out the secret key shared by Alice and Bob. The resulting messaging system is absolutely and totally unhackable—if we believe the quantum textbooks.

In pilot-wave theory the impossible becomes possible—if we have access to subquantum matter and subquantum measurements. If we can watch a particle trajectory without disturbing the quantum wave, in general we can easily tell which quantum wave is guiding the motion of the particle. For example, with wave A the particle might be moving in one direction, while with wave B the particle might be moving in a quite different direction (Figure 93). In that case a subquantum measurement of the trajectory can straightforwardly reveal which quantum wave it really is. This means that, like the uncertainty principle, the problem of distinguishing non-orthogonal states is merely a peculiarity of quantum death. If we escape from quantum death, we *can* reliably distinguish such states and the usual limitation breaks down.

This has two dramatic implications. First, it causes the collapse of the unhackable quantum internet (in the version by Bennet and others). Second, it opens up the possibility of new kinds of computers—which might be even more powerful than quantum computers. Let us turn to that subject now.

Subquantum computers

We should first say a little more about quantum computers—and about computers generally. A computer is basically a device that can store and process data according to given instructions. In abstract terms the basic building block is the 'bit', which can take one of two possible values usually written as '1' and '0'. Data can be expressed as strings of 1's and 0's that encode whatever it is that is being processed. In concrete terms a bit has two distinct physical states. For example, a light switch might be on or off, an electric current might be flowing or not flowing, and so on. In principle, any physical thing with two distinct states can represent an abstract bit of information and so can be employed as a building block for a working computer.

Now, before the advent of quantum mechanics, it was believed that a cat is either dead or alive, a light switch is either on or off, and an electric current is either flowing or not flowing. If we represent the two cases by 1 and 0, then the bit is either in the state '1' or in the state '0'. Obviously. So when the theory of computers was first developed, it was assumed that each bit is in a definite state, reading either 1 or 0. We now call this classical computer science because, as in the classical physics of Newton and Einstein, it is assumed that all physical systems are in definite states.

In contrast, according to quantum mechanics, a cat can be in a superposition of both 'dead' and 'alive'—just as an atom can be in a superposition of two different energy states. This means that a quantum switch can be both on and off, and a quantum electric current can both flow and not flow, at the same time. In other words, a 'quantum bit'—widely known as a qubit (pronounced 'q-bit')—can simultaneously read 1 *and* 0. If we build a computer from qubits, we have something novel called a 'quantum computer'. This device can represent information in different states at the same time.

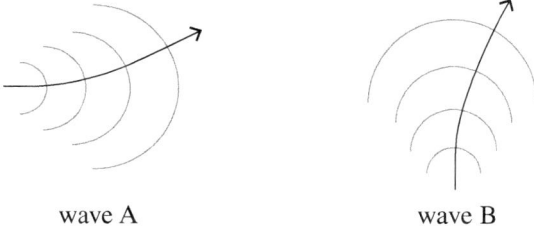

wave A   wave B

Figure 93. Even if quantum mechanics asserts that waves A and B cannot be reliably distinguished for a single particle, in pilot-wave theory the trajectories will generally be different. This means the waves *can* be distinguished by subquantum measurements.

And there is a lot more to it than that. In a quantum computer the qubits change and interact according to the laws of quantum mechanics. This means that qubits can be entangled—in the same way that widely separated particles can be entangled. Different components of the computer can be related in ways that are impossible and unimaginable in classical computer science. In the 1980s, in particular with the pioneering work of Feynman and Deutsch, it started to be widely appreciated that quantum computers can perform tasks that ordinary classical computers cannot.

But are quantum computers really more useful than their classical counterparts? This question was answered with a decisive 'yes' by American mathematician Peter Shor, who in 1994 showed that quantum computers can 'factorise' large numbers much faster than any known classical method. We want to break a number down into its basic building blocks or 'prime factors'. For example, 21 breaks down to 3×7, while 935 breaks down to 5×11×17. It is easy to multiply prime numbers to obtain a bigger number—say if we multiply 5×11×17 to get 935—but it is harder to reverse the process and recover the original numbers. For example, if we are given a six-digit number like 345,567, can we break it down into its prime factors? There are systematic ways ('algorithms') for doing this. The process can be computerised. An ordinary computer can easily factorise a six-digit number in a fraction of a second. But what about a 100- or 200-digit number? The time taken escalates exponentially, so instead of seconds it can take centuries or even millions of years (depending on the size of the number), making it hopelessly impractical. In contrast, for a quantum computer running 'Shor's algorithm', even such large numbers can be factorised quickly, so for example instead of taking a million years it might take only a fraction of a second.[81]

Why would anyone care about fast factorisation? There are in fact important practical implications. Sensitive information sent over the internet, including bank transactions, is currently encrypted by methods whose security depends on the difficulty of factorising large numbers. To hack into the system, we need to be able to factorise large numbers quickly. So far no one has found a way to do that with an ordinary classical computer (a feat that is widely suspected, though not proven, to be impossible). A sufficiently large quantum computer would make it easy—cracking that information wide open to eavesdroppers. In fact, one of the motivations

---

[81] Technically, the time taken grows only 'polynomially' instead of exponentially.

for developing quantum cryptography is to avoid this looming problem that will be created in (probably) the near future by operational quantum computers.

What exactly is it about quantum computers that makes them so much faster? This is an interesting and still-controversial question. In the early days it was thought that the cause of the speed-up was the peculiarly quantum phenomenon of superposition. Intuitively, the idea was that a quantum computer can act like many different classical computers working in parallel—or, as some authors would have it, in parallel universes—where at the end the different computations are brought together to give something useful. This was Deutsch's original point of view, and he regarded the power of quantum computers as evidence for the many-worlds interpretation of quantum mechanics. However, while superposition certainly plays a role, its primacy in explaining the power of quantum computers was eventually disputed. Instead the peculiarly quantum phenomenon of entanglement was highlighted as the cause of the speed-up. Later on, some argued that yet another peculiarly quantum phenomenon, known as 'contextuality',[82] is the true driving force. It seems fair to say there is probably no one simple explanation, and various aspects of quantum mechanics play their part.

In any case, what does pilot-wave theory have to say about all this? Well, first of all, if we are confined to the state of quantum death—so that all of our equipment, including the quantum computer, obeys Born's probability formula—then pilot-wave theory has nothing measurably new to say. It will give the same observable results, just as it does for other quantum phenomena (such as the two-slit experiment). But the physical interpretation is, of course, quite different from that given in the textbooks. According to pilot-wave theory, the remarkable features of quantum computers are explained by the action of the quantum wave, which guides the motion of all the components in a multidimensional configuration space (as for any system of many particles, as we saw at the end of Chapter 2).

Of interest to us now, however, is what happens when we move outside the confines of quantum death. We already know we can then beat the uncertainty principle and send superluminal signals. We also know

---

[82] A quantum measurement is said to be 'contextual' when its outcome depends on what else is being measured at the same time. Quantum nonlocality is an example of this.

we can track the details of hitherto-hidden particle trajectories and exploit that information to undermine the security of quantum cryptography. But would any of this help us build a more powerful computer?

Probably. Recall how tracking trajectories allows us to distinguish different quantum waves that are impossible to distinguish (reliably) in ordinary quantum mechanics (Figure 93). We saw that this can be applied to hack into one version of the quantum internet. It can also form a starting point for building new kinds of computers that are impossible in quantum mechanics. These 'subquantum computers' are likely to be more powerful than their quantum counterparts (though as we will see this is not yet known for certain).

We have said that in quantum mechanics it is impossible to distinguish non-orthogonal states such as 'dead' and 'dead + alive' reliably. Another example would be '1' and '1 + 0'. So imagine we run a quantum computer and one of the qubits ends up with a state (or quantum wave) which we know to be either '1' or '1 + 0'. We will have no reliable way of finding out which it is. So here the theory of quantum computation runs into a wall. Perhaps the state of that qubit encodes some useful information, which we can find out only if we can tell whether the state is '1' or '1 + 0'. But quantum mechanics says we cannot. So there might be some crucial information there inside the computer, locked into the quantum wave of one of the qubits, but we cannot read it.

Unless we happen to be outside the confines of quantum death. In that case, as we have seen, pilot-wave theory allows us to track trajectories and distinguish the quantum waves '1' and '1 + 0'—allowing us to unlock information that is normally hidden. So there we have it. The wider physics of pilot-wave theory allows us to distinguish non-orthogonal states, contravening a basic theorem of quantum mechanics, and by this means we can extract more information from a quantum computer than would normally be possible. We might then consider building a subquantum computer with a component that implements this ability (Figure 94). Such a device could certainly perform tasks that are impossible for any quantum computer. But would this make the computer more powerful?

This question was first studied in a different context (not involving pilot-wave theory). It has sometimes been suggested that the quantum wave might change with time in ways not normally allowed by the Schrödinger equation. What is known as 'nonlinearity' would allow non-orthogonal states to be distinguished. We emphasise that this has nothing to do

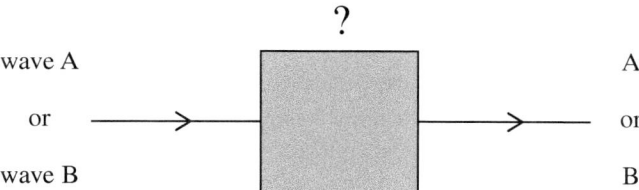

wave A   A

or   or

wave B   B

Figure 94. A computer component (or 'gate') that can reliably distinguish non-orthogonal quantum states 'A' and 'B' would allow a subquantum computer to perform tasks that are impossible for conventional quantum computers.

with pilot-wave theory or quantum nonequilibrium.[83] In 1998 American physicists Daniel Abrams and Seth Lloyd looked at how powerful this capability would be if implemented on a quantum computer. They constructed a quantum algorithm in which the answer to a crucial problem is encoded in the small difference between two non-orthogonal states. In other words, to extract the answer from the computer we would need to distinguish two quantum waves that are normally impossible to distinguish. Abrams and Lloyd showed that, if those two waves could be distinguished, we would have a much more powerful computer.[84] They suggested this might be possible in practice if one day we find 'nonlinear terms' in the Schrödinger equation (such terms have been searched for experimentally but none have been found yet).

Now let us return to pilot-wave theory. If we are outside the confines of quantum death, we can distinguish the two quantum waves that appear at the end of the algorithm constructed by Abrams and Lloyd (and we can do this with no need for nonlinearity). We then seem to have a clear conclusion: the new nonequilibrium physics of pilot-wave theory allows new kinds of computers that are even more powerful than quantum computers.

But, in fact, there are some subtleties which have not been sorted out yet. The new physics of pilot-wave theory certainly allows a new kind of computer, in which the answer to a hard problem is found from subquantum measurements that distinguish between non-orthogonal states. That much is clear. But to evaluate the power of a computer we need to

[83] Technically, nonlinear terms in the Schrödinger equation can allow initial non-orthogonal states to evolve into final orthogonal states (which can then be distinguished as usual).

[84] Technically, the computer could solve 'NP-complete problems' in polynomial time.

know how quickly the required 'resources' grow with the size of the task. If the resources grow exponentially with the task size then the algorithm is regarded as inefficient, while if they grow only polynomially then the algorithm is regarded as efficient. So how powerful is a subquantum computer? To answer this we need to consider the resources—the subquantum matter—required to carry out the subquantum measurements. How much subquantum matter will we need, and how far from quantum death (or quantum equilibrium) will it have to be? And, crucially, how rapidly will these resources have to scale up with the size of the computational task? At the time of writing these questions have not been answered. And so the true power of subquantum computers remains an open question for future research.

Interplanetary communication

Finding a way out of quantum death would certainly herald a revolution in information technology and computing. No doubt there would be many other applications too that we are presently unable to think of. But let us now return to the simplest and most obvious application: instantaneous or nonlocal signalling (discussed in Chapter 5). What would it be good for?

A radio message sent from London to New York at the speed of light takes about one-fiftieth of a second to arrive. No one would get too excited if we could send the same message faster or even instantaneously. What practical difference would it make? But now there is talk of building a permanent human base on the Moon. It would take about a second for our friends up there to receive a radio message from Earth. And it would take another second for us to receive their reply. Real-time conversation would still be easy enough, there would just be an annoying delay between speaking and hearing, as sometimes happens with a faulty telephone line. But we get into different terrain if we start thinking about interplanetary exploration. If we build a human base on Mars, as some are planning to do, then for a radio signal to travel each way will take up to about 20 minutes (depending on the current positions of Earth and Mars). So imagine the stilted conversation: we speak and it takes 20 minutes to be heard at the other end, and then there will be another 20 minutes before we hear their reply. This is already an issue with robotic landers exploring the surface of Mars. Radio control and directions from Earth cannot be given in real time. The delay is too long. And should the lander run into trouble, it will be up to 20 minutes before we know about it. If we venture further out into the Solar System, these numbers get worse. It takes about four hours for a radio signal from Earth to reach the planet Neptune.

So here is a first and obvious application of 'subquantum technology': practical interplanetary communication in real time. As we saw in detail in Chapter 5, superluminal signalling is possible only if we have matter in a nonequilibrium state that violates Born's formula. If we had a supply of such particles we could harness them for instantaneous communication over large distances. But there is a catch: the particles have to be entangled. So imagine we want to signal to our friends at the base on Mars. We need a supply of nonequilibrium (or subquantum) particles at both ends—*and* those particles have to be entangled. How can this work?

Entangled photons are easily created in the lab when pairs of photons are emitted by a single atom.[85] More generally, if we let two particles interact in some way they will become entangled. We can then pull them far apart, as far away from each other as we like, and they will remain entangled. The spooky connection continues even across the distant reaches of space (for as long as the entangled quantum wave is not disrupted by interactions

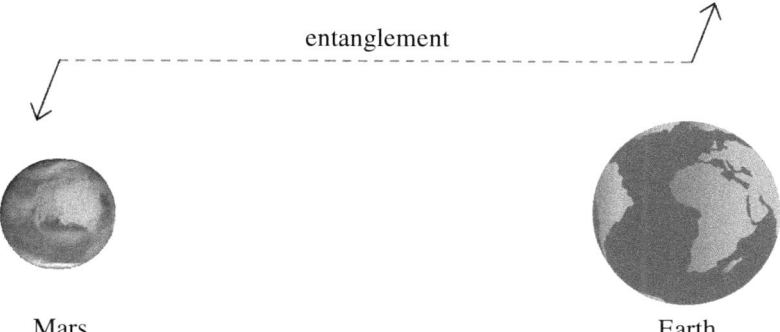

Figure 95. A line of instantaneous communication between Earth and Mars requires entangled particles to be shared between the two planets.

---

[85] In a two-step cascade emission (Chapter 9).

with other particles[86]). But to generate the entanglement in the first place, we need the particles to start off close together. This means that, to set up our Earth–Mars Instantaneous Communication System, we will have to transmit nonequilibrium particles to Mars that have already been entangled with nonequilibrium particles kept here on Earth (Figure 95). The particles might be transported by spacecraft. During transit they would need to be carefully shielded from interactions with other particles, to avoid degrading the entanglement, and also to avoid relaxation. Alternatively, the particles might be fired from Earth towards a receiving base on Mars—perhaps as a laser beam, if the particles are photons. There would certainly be practical difficulties to overcome. But there is no reason of principle why nonequilibrium entangled particles could not be shared between the two planets, thereby setting up a line of instantaneous communication.

In practice, however, it might be easier to set up the entanglement link using ordinary equilibrium particles, which are readily available and for which there is no issue about avoiding relaxation. For such particles, any underlying instantaneous signals will be obscured by quantum noise. However, if we have a supply of (unentangled) nonequilibrium particles on each side, these can be used to perform subquantum measurements that beat the uncertainty principle (Chapter 5). An experimenter on Mars could then monitor the precise motion of a single entangled particle, without disturbing the quantum wave, and so detect instantaneous signals from Earth. Similarly, to receive a reply on Earth, we will need to perform subquantum measurements implemented with a local supply of nonequilibrium particles.

Obviously, should subquantum matter be discovered, there will be engineering challenges involved in its storage, transport, and practical use (in particular, to avoid relaxation to the usual state of quantum death). These will probably be comparable to the efforts that are now being made to build quantum computers, where delicate qubits need to be carefully shielded from outside influences. But if the rewards are worth it, the effort will be made. So one day, possibly, subquantum matter will be deployed for instantaneous, real-time, interplanetary communication across our solar system. And in principle there is no reason why this could not be extended to even greater distances beyond. This brings us to our final topic.

---

[86] Technically, interactions can cause 'decoherence' and an effective collapse of the quantum wave, which can destroy the entanglement.

Is anyone out there?

For centuries humanity looked out across the reaches of space and saw 'the heavens'—an eternal ethereal realm beyond earthly comprehension, inaccessible to communication (except perhaps through prayer) or to travel (except perhaps in the afterlife). Today we look out into space and see vast stretches of interstellar and intergalactic space extending to billions of light years—with present technology still inaccessible to communication or to travel. Our universe is populated by billions of galaxies, each containing billions of stars. Most of those stars have orbiting planets. Inevitably, we find ourselves wondering about intelligent life on other worlds, and the possibility of some form of contact.

The nearest star, Proxima Centauri, is about four light years away. It has an approximately Earth-sized planet (Proxima Centauri b) orbiting in the 'habitable zone'—neither too close to, nor too far from, the star, so the ambient temperature allows the presence of liquid water. The planet was discovered in 2016 and widely hailed by astronomers as an ideal candidate for future interstellar exploration and the search for life beyond Earth. Whether or not the planet harbours life, intelligent or otherwise, is not yet known. In any case, to send a radio message to our potential nearest neighbours would take four years—with the message crawling across space, as a tiny ripple in the electromagnetic field, at the absurdly slow speed of light. To receive a reply would take another four years. Many other 'exoplanets'—planets beyond our solar system—have been found. For example, several have been seen orbiting the star HR 8799 about 130 light years away. Other planets have been found at distances of thousands of light years. Looking towards the centre of our galaxy, news takes some 26,000 years to travel by radio. To other galaxies it takes millions or even billions of years. If there is intelligent life elsewhere in the universe, our present technology offers little chance of conversation. We are trapped and bound within a tiny sphere, almost as the ancients were in their self-imposed Ptolemaic system.

How likely is intelligent life—or indeed any life at all—elsewhere? A strong case for life outside Earth has been building for at least half a century, beginning with the stunning discovery that the chemical ingredients for life as we know it are to be found throughout the universe.

In 1969 the Murchison meteorite landed in Australia. It was a large rock from outer space, which broke into fragments as it fell through the atmosphere. Astonishingly, it was found to be rich in organic (carbon-based) compounds, including some dozen amino acids, which are generally seen as key components of living organisms. Around the same time the organic compound formaldehyde was discovered in interstellar space.[87] Since then a huge number of complex organic molecules have been identified not only in nearby interstellar clouds but across our galaxy as well as in other galaxies.

The detected organics include 'prebiotic' molecules commonly associated with the origin of life. Remarkably, recent analysis of the Murchison and other meteorites has shown the presence of all the nucleobases needed to build the genetic code as we know it, including the base pairs that in living organisms connect the two strands of the DNA double helix. Extraordinary as it may seem, the basic building blocks of organic life are widespread across the universe.

Astrochemists have developed an understanding of how these organic molecules were formed, by comparing with molecules produced in laboratory simulations of interstellar space. Recent experiments show that a mixture of water, carbon monoxide, ammonia, and methanol (common components of 'interstellar ice'), kept at 10 degrees above absolute zero and irradiated by intense ultraviolet radiation (to mimic radiation from star formation), over time produces organic molecules similar to those found in space—including most of the nucleobases found in DNA and RNA on Earth. The wider universe is, in effect, a gigantic organic chemistry lab, in which complex organic and even prebiotic molecules are being produced on an industrial scale.

Contrary to what had once been thought, our universe is a dirty and smelly place. It also contains vast quantities of water (mostly in the form of vapour or ice). Water has been found in interstellar clouds, in material swirling around stars forming new planetary systems, and in the atmospheres of remote planets orbiting other stars. Our solar system contains staggering amounts of water. Beneath the icy surface of Jupiter's moon Europa, there is evidence for a saltwater ocean up to 100 miles deep (kept liquid by tidal heating effects). An underground ocean is also thought to exist beneath the icy surface of Saturn's moon Enceladus, feeding jets of ice grains and vapour that spray visibly from cracks in the surface and out into space.

---

[87] Specific molecules in space can be identified by the frequencies of radiation which they absorb or emit.

In 2018 a suite of organic molecules was detected in the jets erupting from the depths of Enceladus, suggesting complex organic chemistry taking place within the underground ocean. Could it harbour living organisms? If this seems an unlikely place to look for life recall that, even on Earth, 'extremophile' organisms live in conditions of intense heat and cold, high pressure, and complete darkness. Simple organisms have been found in deep-sea volcanic (hydrothermal) vents at temperatures exceeding 80 degrees Celsius. Others live deep underground, with energy derived from natural radioactivity in rocks. These discoveries suggest that life could exist under similar extreme conditions in our solar system and beyond.

The Solar System formed about 4.5 billion years ago from the collapse of an interstellar cloud of gas and dust. It is widely thought that the early Earth was enriched with prebiotic chemicals during the 'late heavy bombardment' by asteroids from space. Meteorites and icy comets also delivered large quantities of water. That chemical enrichment may well have contributed to the development of life on our planet, even if there are still gaps in our understanding of how the first self-replicating biomolecules arose.

There is some evidence that the first microorganisms may have arisen on our planet as early as a few hundred million years after its formation. This suggests that the development of simple life forms is easier than was previously thought—a natural consequence of basic physics and chemistry, in a sense as natural as the formation of atoms and molecules, or of stars and planets. It increasingly seems just a matter of time before we discover life elsewhere—if only in the form of simple microbes.

Some have argued that, while simple organisms are likely to be abundant, more advanced multicellular animal- or plant-like life is probably exceedingly rare and perhaps even unique to Earth. These sceptics point to apparently rare characteristics of Earth, such as having a magnetic field to shield us from the solar wind, an unusually large moon (which stabilises the Earth's tilt and so regulates the climate), and plate tectonics. Critics counter that all planets are likely to have their idiosyncrasies, and that none of these features have been proven to be essential for the development of higher life forms. In any case, recent evidence suggests that these features may not be so rare after all. Finally, tiny multicellular animal life has in fact already been found in extreme environments (for example, up to two miles below the Earth's surface) that are likely to be replicated elsewhere in our solar system and beyond.

The search for extraterrestrial intelligence

The possibility of complex extraterrestrial life raises the question of whether there might be advanced scientific and technological civilisations elsewhere in the universe. In only a few thousand years we progressed from the wheel to the steam engine, and in a mere few hundred years we made the leap from the steam engine to interplanetary spacecraft. We know that the Galaxy is about nine billion years older than the Sun. It contains Sun-like stars that formed long before our own star did. If Earth-like conditions exist on planets elsewhere, in many cases any intelligent life that may have developed there will have had a head start of millions or even billions of years. Thinking this through it becomes evident that, if we are not alone, the Galaxy probably contains civilisations that are immeasurably more advanced than ours. This raises a host of difficult questions—not least of how alien technology might compare with ours, and what meaningful contact might look like.

Are there intelligent aliens? For whatever reason, the question tends to provoke intense scepticism and unrestrained gullibility in equal measure (rather like the search for the paranormal). For thoughtful observers, it is unclear which conclusion would be the more troubling: the discovery of intelligent aliens, or the discovery that we are alone in the universe.

Whatever the answer may be, some effort has been made to signal our own presence to any aliens that may or may not be out there. It has long been known, by theoretical calculation, that our radio technology is powerful enough to permit interstellar communication (at the speed of light). In 1975 the Arecibo message—containing basic biological and astronomical information about humanity and Earth—was broadcast from a radio telescope in Puerto Rico and directed towards a star cluster about 25,000 light years away. Needless to say, as yet there has been no reply. More concrete means of signalling our presence have also been tried. The two *Voyager* space probes, launched by NASA in 1977 to explore the outer Solar System, each carry a gold-plated disc with images and sounds from Earth and its peoples. Those probes are now in interstellar space and will continue their journey through the Galaxy. In about 20,000 years they will approach the nearest star. If the probes are eventually intercepted by spacefaring aliens, the immortal music of Mozart may find a new and appreciative audience.

Much more effort has been devoted to the search for signals *from* aliens. With humble beginnings in the early twentieth century, the Search for

Extraterrestrial Intelligence (SETI) is now a well-funded scientific enterprise. Since the 1960s various astronomical programmes have scanned the sky for radio signals from deep space, looking for signs of communications from alien civilisations in our galaxy and beyond. The latest and largest such programme, Breakthrough Listen, initially based at the University of California, Berkeley and now headquartered at the University of Oxford, started observations in 2016 and aims to 'listen' to one million nearby stars as well as the centres of 100 nearby galaxies. Radio emission is a natural phenomenon in the universe (for example, from charged particles swirling in magnetic fields) and 'radio astronomy' has long been a mature science. The point of SETI programmes is to see if radio emissions from space might contain some kind of pattern or regularity that would not be produced by natural sources and which might encode a message or communication from an alien intelligence. So far, despite extensive searches, no such signals have been found.

The silence is arguably deafening. If the Galaxy is teeming with advanced intelligence, why is the radio sky not abuzz with alien communications? This question is part of a larger problem called the 'Fermi paradox', named after the Italian physicist Enrico Fermi, who famously raised it during a lunchtime conversation with fellow scientists around 1950. Since Fermi's time the paradox has been much studied and sharpened. For example, once a civilisation masters the art of interstellar travel—and we are ourselves not far off from that—we might reasonably expect it to gradually colonise the Galaxy. It has been estimated that this would take one or two million years—a long time by our standards but short compared to the age of the Galaxy. And so the Galaxy should have been overrun, long ago and repeatedly, by advanced aliens. This raises the question, as Fermi reportedly put it: 'Where is everybody?'

Despite plausible expectations, we see no obvious signs of alien colonisation—certainly not in our solar system. And despite decades of searching, our radio telescopes find no indication of alien communications from other planetary systems. This is the essence of the puzzle: if they exist, why do we not see them?

This is a difficult question and many attempts have been made to answer it. For example, it has been suggested that technological civilisations quickly destroy themselves (as we might soon do), or that advanced aliens inhabit electronic computer simulations and have no need to spread physically over space, or that our cosmic habitat is so dangerously predatory as to discourage any advanced life from advertising its own existence.

Others argue, as noted, that higher (multicellular) life forms are likely to be exceedingly rare in our universe, and that we might be alone at least in our galaxy.

The Fermi paradox is a serious and multifaceted question, and there have been far more ingenious attempts to answer it than we can summarise here.[88] There is much disagreement in the field—for example, about the likelihood of aliens wishing to make contact. But on one point almost all commentators agree: intelligent aliens, if they exist, will be bound by the limitations of our own presently known laws of physics. And it is here that we part company with mainstream thinking on the subject.

The question of alien technology

Most astronomers searching for extraterrestrial intelligence assume that aliens will communicate by making small ripples in the electromagnetic field which spread across space at the speed of light. These ripples, otherwise known as radio waves, are generated by shaking charged particles up and down (Figure 96). More formally, a radio wave is generated by an 'oscillating current' in the transmitter. The wave moves across space and eventually shakes charged particles up and down over there, generating an oscillating current in the receiver. In this way a signal is sent from here to there at the speed of light. That is how radio transmissions work. This technology was discovered by Earthlings in the late nineteenth century and

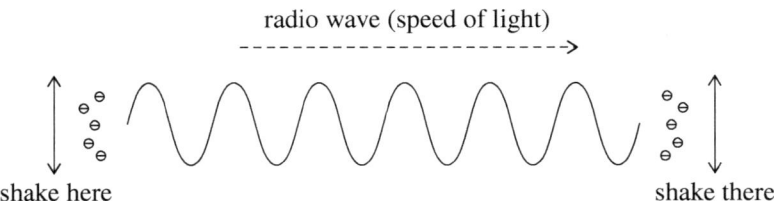

radio wave (speed of light)

shake here   shake there

Figure 96. Is this an advanced form of communication? By shaking charged particles up and down over here, we can shake charged particles up and down over there—with a delay set by the speed of light, as ripples in the electromagnetic field make their way from here to there.

[88] Stephen Webb's book *If the Universe is Teeming with Aliens . . . Where is Everybody?* discusses no less than seventy-five possible answers.

water wave (one metre per second)

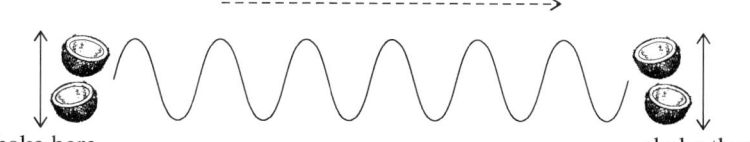

shake here   shake there

Figure 97. Is this an advanced form of communication? By shaking coconut shells up and down over here, we can shake coconut shells up and down over there—with a delay set by the speed of about one metre per second, as ripples in the water make their way from here to there.

was put to widespread use in the twentieth. Many Earthlings still believe that this is an advanced form of communication. But is it?

The 'People of Earth' live on the edge of a lake in a remote region of the Amazon rainforest. They have never had contact with the outside world. Long ago they split into two groups living on opposite sides of the lake. They cannot see each other directly because the lake is usually covered in mist. So they use water waves to communicate with their fellows on the other side. It is quite ingenious. They drop some coconut shells into the water and give them a good shake up and down to generate a rippling pulse, or water wave, which spreads out at a speed of about one metre per second and soon reaches the opposite shore (Figure 97). On the other side they are ready. They have their own coconut shells floating in the water, which noticeably bob up and down when the wave signal strikes. They take turns monitoring the floating shells and watching for signals. Despite their simple conditions, the People of Earth have mastered the art of coding. One long rippling pulse means '1', a shorter pulse means '0'. So on the other side they might receive a message such as '101100110010110' which means 'please send us more fish'. One day their Chief Coder—an introspective type, given to taking long solitary walks—begins to wonder if there might be other Peoples living elsewhere on the lake. 'Where is everybody?' he exclaims one day over lunch. Of course the lake is never perfectly still. There are always some waves here and there. That is just because of the wind. There is a clear difference between the random pattern of the natural waves and the distinctively structured waves arriving as signals from the opposite shore. But what if they are missing something? And so begins their quest for contact with other civilisations beyond their own. The Chief Coder leads a team to monitor their floating coconut shells for unexpected patterns in the seemingly random and natural water waves. But years of searching turn up nothing. And all the while, streaming invisibly through the air around them, are radio waves from satellite television channels broadcasting programmes about climate change and the importance of preserving the Amazon rainforest—programmes that are watched by millions around the world who happen to possess a television set capable of picking up those signals and converting them into an audiovisual experience.

Is our situation any different? Shaking charged particles up and down to generate an electromagnetic wave is certainly a step up from shaking coconut shells up and down to generate a water wave. As a means of communication it is faster and more efficient. Even so, it is the same *kind* of thing. Remember in Chapter 2 when we talked about the idea of 'local action': something here wiggles and causes an effect close by, which in turn wiggles and causes another effect a bit further away, and so on. That is how a water wave spreads out. That is how a sound wave spreads out. And that is how an electromagnetic wave spreads out. Like a domino effect the original pulse moves step by step across space at a finite speed. Our physics textbooks tell us that all known interactions take place in this way. In that sense we have progressed little beyond the water wave. We still use local action to communicate across space. And when we begin to wonder about extraterrestrial intelligence, we assume as a matter of course that any advanced aliens out there will *also* use local action to communicate across space—specifically in the form of electromagnetic waves. Might it be that, instead, streaming invisibly through the space around us, there are signals generated by a more advanced technology which we simply lack the understanding and wherewithal to tap into?

After his ashes were interred at Westminster Abbey in London in 2018, a radio message from Stephen Hawking was broadcast into space from a satellite dish in Spain operated by the European Space Agency. The beam was aimed towards the nearest known black hole (in the constellation Monoceros)—a touching gesture and a fitting tribute to a physicist who had spent much of his life deepening our understanding of these enigmatic objects. And yet the message is a small ripple in the electromagnetic field, crawling across space at a speed which appears tiny on a cosmic scale. It will take more than 3,000 years to reach its destination. Is there not a better way to communicate across the universe?

Some astronomers have suggested that laser beams might be more effective than radio waves, and indeed searches have been made for extraterrestrial laser signals from nearby stars in our galaxy. Ongoing searches for laser communications are also a key part of the Breakthrough Listen programme. But, again, any such signals will be limited by the usual speed of light.

American-British theorist Arjun Berera and German astrophysicist Michael Hippke have recently suggested that aliens might be communicating by means of specifically quantum properties of electromagnetic radiation—such as the polarisation states of single photons. Well-known techniques of 'quantum communication' have certain advantages in terms of signalling efficiency. And, as we have seen in this chapter, quantum cryptography also provides additional security (as long as we are confined to quantum death). Berera and Hippke have in effect suggested that something like the quantum internet, which humanity is constructing on Earth, might already exist in interstellar space (where aliens may have deployed spacecraft to share entangled particles across large distances, or simply fired beams of such particles across space). Perhaps programmes such as Breakthrough Listen should extend their remit to include searches for specifically quantum signatures in radiation from deep space. This is all fair enough. However, while these quantum techniques are certainly more sophisticated than signalling with radio waves, the speed of transmission is still limited by the speed of light. An alien message or quantum communication from a planet 500 light years distant will still take 500 years to reach us. So again we must ask: is there not a better way to communicate across the universe?

If there is a better way, advanced alien intelligence may have found it. In this book we have seen how subquantum matter—nonequilibrium particles that have escaped quantum death—can be deployed to send instantaneous signals. And we have considered how we might one day make use of this for interplanetary communication. Could an extraterrestrial intelligence employ the same effects to communicate instantaneously across the vast reaches of interstellar and possibly even intergalactic space? If so, current astronomical programmes searching for extraterrestrial intelligence will prove to be as misguided as searching for intelligent life on Earth in the twenty-first century by monitoring for signals in water waves.

Communication across the universe

Let us assume that intelligent aliens—on a remote exoplanet—have discovered subquantum matter and are able to put it to practical use. Perhaps they have harvested nonequilibrium particles from exploding primordial black holes, or otherwise collected nonequilibrium relics from the early universe, or maybe they are able to create such particles by some means unknown to us. How might they harness this resource to communicate across the universe?

To communicate with us, or with some other distant target audience, instead of sending a conventional radio or laser pulse, they might send a pulse of nonequilibrium particles that are entangled with other nonequilibrium particles kept at the alien source. As far as we know, the particles will necessarily travel at a speed no greater than the speed of light. But once received, the particles will open a line of instantaneous communication between the source and the receiver—just as we discussed between Earth and Mars (Figure 95). The precise motions of the received particles will depend visibly and instantaneously on what is done to their entangled partners at the source (Chapter 5). Closely monitoring the received particles can then reveal instantaneous messages. And the communication channel can work both ways: local forces applied to our particles will have an instantaneous effect on the motion of distant particles at the source, enabling us to send messages in reply.

As in the case of Earth–Mars communication, in practice it might be easier to establish an entanglement link using ordinary equilibrium particles, which are much more abundant and for which there is no issue about avoiding relaxation. The obscuring effects of quantum noise could then be circumvented, however, only by means of subquantum measurements—assuming we have a local supply of subquantum matter with which to implement such measurements. By monitoring the precise motions of the incoming particles, without disturbing the quantum wave, we will be able to detect instantaneous signals from the distant source. In other words, aliens might send us pulses of equilibrium particles, which are entangled with other equilibrium particles kept at the source, and we will be able to read incoming instantaneous messages only if we are able to perform subquantum measurements that beat the uncertainty principle (Figure 98). By cutting through the fog of quantum noise that permeates our universe, we might reveal real-time signals from distant alien civilisations.

In the search for extraterrestrial intelligence, a new tactic could be to monitor the sky for incoming subquantum particles—photons and other particles whose quantum states break Born's formula. As we have seen, deviations from Born's formula could show up as a modified wavy pattern

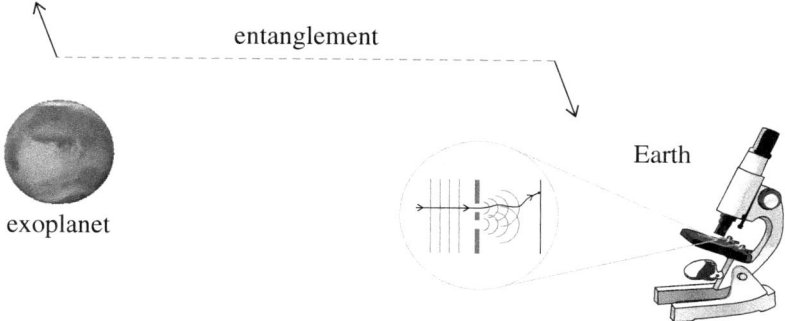

Figure 98. Instantaneous signals from a remote exoplanet. If the incoming particles obey Born's formula, to read instantaneous messages from the source we will have to perform subquantum measurements which beat the uncertainty principle.

in a two-slit experiment. Alternatively, we might look for anomalies in polarisation probabilities. Should such particles be found, and should they originate (for example) from a remote exoplanetary system, we might do well to monitor them for possible incoming alien communications.

We have said that spacefaring aliens, with an evolutionary head start of millions or even billions of years, could have already colonised the Galaxy. This means that, instead of firing pulses of entangled particles at desired targets, they could transport such particles with them as they travel, thereby setting up instantaneous lines of communication across the Galaxy. However, whether the particles are transported by spacecraft or fired in the desired direction, as far as we know the speed-of-light travel barrier will apply. So the lines of communication can be extended no faster than the speed of light. If we are talking about distances of hundreds, or perhaps thousands, of light years (within the Galaxy), great patience will be required before the desired instantaneous communication channels can be activated.

There may be a better way. Perhaps nature herself has already done the work, by creating entangled pairs of particles in the early universe. As the particles in each pair move apart—either because they are moving away from each other, or because space itself is expanding—they can remain entangled even when they are separated by vast distances. The upshot is that relic particles from the early universe can *already* be entangled across huge distances, providing a naturally occurring resource for instantaneous communication. Even if the relic particles have now reached equilibrium, they can still provide a channel for instantaneous communication, provided the recipient of the message has a supply of subquantum matter with which to circumvent the uncertainty principle.

There are at least two scenarios whereby 'entangled relics' can be created in the early universe. We can describe them briefly.

First, as we saw in Chapter 7, according to inflationary cosmology, the matter in our universe was created at early times by a process of inflaton decay, whereby the energy of the inflaton field was converted into matter and radiation. Entangled pairs of particles can be created by this process—just as entangled pairs can be created by the decay of particles in experiments at high-energy accelerators. If the early inflaton field carries nonequilibrium, so will the decay products.

Second, in the early universe, entangled pairs of particles can be created out of the vacuum by the rapid expansion of space. This may sound mysterious, but it is a well-known consequence of applying quantum mechanics to certain kinds of fields in an expanding universe, and it can also be described by pilot-wave theory. In effect this process creates entangled pairs of particles, where the particles within each pair move in opposite directions.[89] If the initial vacuum fields (associated with seemingly empty space) break Born's formula, so will the created pairs.

On either scenario, whether they were created by inflaton decay or by the expansion of space, entangled relics from the early universe could provide a ready-made resource for instantaneous signalling today. For all we know this resource may already have been exploited by advanced aliens, and it might one day be harnessed by us.

There are, however, important caveats to consider. A delicate web of entanglement across large distances in space can become degraded by interactions with other particles. Interstellar and intergalactic space is, for the most part, close to a vacuum, but even so some interactions will occur and this must be taken into account. The effectiveness of entanglement—as quantified by the 'fidelity' of quantum communications—can also be disrupted by gravitational disturbances from bulk matter.

These complications have been studied by Berera and collaborators (who, it should be said, do not consider violations of Born's formula

---

[89] Technically, this process cannot occur for the electromagnetic field owing to 'conformal invariance'. So any entangled pairs created by the expansion of space will consist not of photons but of other particle species.

or superluminal signalling, but focus on quantum communication at the speed of light). The most recent calculations, by Berera and Ecuadorian theorist Jaime Calderón-Figueroa, indicate that pairs of photons can remain entangled across interstellar distances (within the Galaxy). The details depend on the wavelength, with different results for X-rays, radio waves, or optical photons. Furthermore, it appears that gravitational disruption can in principle be mitigated if we know the path travelled from source to receiver. If all this is correct it means that, in practice, entangled photons could provide a resource for instantaneous communication over at least hundreds and perhaps thousands of light years.

We might ask if this could be extended to intergalactic distances— hundreds of thousands, millions, or even billions, of light years. In fact, earlier calculations by Berera and collaborators show that interactions are quite small for photons traversing intergalactic space, raising the possibility of entanglement persisting between Earth and distant galaxies. But the extent to which this might enable practical communication needs further study.

Beyond its practical applications, instantaneous signalling across space will also have theoretical implications for our understanding of time. As we discussed in Chapter 4, instantaneous signals could be employed to synchronise distant clocks and to define an absolute time across the universe. This can work even when gravity plays a significant role and spacetime is curved.[90] Contrary to Einstein's interpretation of relativity physics, there is an underlying true time and a true state of rest, though these can be discerned only outside the confines of quantum death.

On our current understanding, then, a web of entangled particles could exist in our universe, with entanglement persisting over interstellar and perhaps even intergalactic distances. The particles may have been distributed by alien intelligence, or they might be natural relics from the Big Bang. According to pilot-wave theory, such particles can provide a channel for instantaneous communication (Figure 99). If the particles break Born's formula, they will in themselves enable instantaneous signalling. If instead they have reached equilibrium, we will need a supply of uncertainty-violating subquantum matter to be able to read any

---

[90] Technically, as noted in Chapter 4, an absolute time can be defined when spacetime is globally hyperbolic (widely regarded as an essential condition on physical spacetimes).

Figure 99. Communication across the universe. Space may contain a web of entangled particles, distributed by alien intelligence or arising naturally as relics from the Big Bang, providing a channel for instantaneous signalling (dashed lines). (Artist's illustration of exoplanet OGLE-2005-BLG-390Lb, about 20,000 light years from Earth.)

incoming instantaneous messages. In either case, there will doubtless be further practical hurdles to overcome in attempts to harness such particles for instantaneous communication. For us these are problems for future research—problems which may have already been solved by advanced alien intelligence.

If our fanciful People of Earth, isolated in the Amazon and communicating by means of water waves, managed to construct a simple radio receiver, they would discover to their astonishment that our world is abuzz with electromagnetic communications travelling at the speed of light. Should humanity eventually discover a web of entangled particles in space, and successfully monitor them to a precision better than is allowed by the uncertainty principle, will we discover that our universe is teeming with instantaneous communications from advanced alien civilisations? Possibly. At the very least, the harnessing of instantaneous signals across space would provide us with a new window onto the distant universe—as well as a window onto quantum reality itself.

Epilogue

The edge of the world

Since antiquity our scientific knowledge has advanced in fits and starts. Sometimes we have spent centuries lost down blind alleys, other times we have progressed in leaps and bounds. For some 200 years, roughly from 1700 to 1900, it was widely believed that Newton had finally uncovered the true laws of nature. Then cracks appeared, followed in quick succession by two earthquakes—the relativity and quantum revolutions. By 1927 it was all over, or so it seemed. Almost right away there emerged a widespread belief that we had reached the end of the road. While Einstein thought that his new theory of gravity was only a stepping stone to something deeper, Heisenberg believed that quantum mechanics was final and definitive ('endgültig'). Remarkably, even though we had recently emerged from 200 years of believing that we had the finally valid physics, which had just been overthrown not once but twice, it was claimed that *now* we had the ultimate answer.

We had in effect reached the edge of the world. Like archaic sailors on remote voyages, we had better stop for fear of falling off into the infinite void. Roman and Medieval cartographers at least wrote *Hic sunt leones*— 'Here are lions'—at the edge of their maps where unknown territories remained to be explored. But at the edge of our most modern map—our most fundamental theory of the world—there is a sign saying: 'here is nothing'. Even to *think* beyond the edge is 'meaningless'. Might there be something wrong with this?

The boldness of Heisenberg's claim, and its contrast with Einstein's more modest stance, was pointed out by the Austrian-British philosopher Karl Popper in his perceptive book *Quantum Theory and the Schism in Physics* (published in 1982). He called Heisenberg's attitude the 'end-of-the-road thesis'. As Popper noted:

Beyond the Quantum. Antony Valentini, Oxford University Press. © Antony Valentini (2025). DOI: 10.1093/oso/9780198853749.003.0012

Quantum mechanics was to be regarded as the Last Revolution in physics, since it had reached the inherent limits of knowledge.

Why did Heisenberg (and Bohr) believe that we had reached the 'inherent limits of knowledge'? What does that even mean? Throughout history there has been a recurring belief that we had finally understood everything—perhaps it is only human to hope that we are fortunate enough to be born into a time when all has been revealed. But this time there was something different, something deeper, something more damningly final: this really was the end of the road, beyond which deeper knowledge was impossible. Why?

The ultimate limits of human knowledge

There are two strands to the argument.

The first rests on a philosophical interpretation of the uncertainty principle: we are unable to observe the details of the microworld; therefore it makes no sense to talk about them—they are meaningless. What we cannot measure cannot be spoken of, and so does not exist. This is of course a modern version of the age-old conflict between perception and reality, though this time the supporters of perception are claiming a complete and *final* victory.

The laws of physics themselves—at least as interpreted by Bohr and Heisenberg—dictate that we have reached the end and can go no further. We are immersed in a sort of ultimate impenetrable quantum fog that is mandated by the laws of the universe. There is an obvious riposte. What if those laws eventually turn out to be wrong or limited? Are we not *assuming* that those laws are final, in order to draw the conclusion that the consequences of those laws are final?

The second strand of the argument is more overtly philosophical: we have reached the limits of what the human mind can understand. Our minds are able to comprehend the macroscopic world of tables and chairs, billiard balls and planets, but we simply lack the mental wherewithal to comprehend the microscopic world of atoms and elementary particles. Why would anyone believe that? On this point Bohr and Heisenberg were deeply influenced by the eighteenth-century German philosopher Immanuel Kant.

Kant's writings are notoriously obscure and their meaning remains controversial to this day. But historically they have been widely interpreted as follows. The physical world as we see it is shaped by the structure of the human mind. For example, when we see a tennis ball moving through space, we might assume that we are seeing reality as it is, but according to Kant (as widely interpreted) the very idea of a 'material object' moving through 'space' is really just that—an idea, a way of thinking, which our minds impose on the world. And what is more, we have no choice in the matter. We *have to* think about the physical world using the concepts and laws of Newtonian physics, not because those concepts and laws describe reality, but because they are part and parcel of the unalterable fabric of human thought. On this view, Newtonian physics is not so much a map of objective reality but a map of the structure of the human mind. These ideas first appeared in Kant's famous *Critique of Pure Reason*, published in 1781. It was without doubt one of the most influential philosophical books ever written.

Let us now fast forward to the early twentieth century. It was found that Newton's physics did not work inside atoms. What to do? According to Kant, without Newton's physics we cannot think clearly about the physical world. If we accept this we are driven to an extraordinary conclusion: we are, by our very nature, unable to think clearly about what is happening inside atoms. And so the human mind can never truly understand atomic physics. Our minds are able to cope with the everyday macroscopic world of classical physics, but we are out of our depth in the microworld and forever unable to understand it—not because there is something we do not know, but because there is something we cannot ever fathom. We have reached the ultimate limits of human knowledge and understanding.

Kant is widely regarded—especially in continental Europe—as one of the greatest philosophers who ever lived. So now deep philosophy also dictates that we have reached the end and can go no further. Again there is an obvious riposte: what if Kant was wrong or in some way limited? After all, Kant also claimed that Euclidean geometry—the simple geometry of triangles and circles that we learn at school—was a fixed feature of the human mind, and that we could not think about geometry in any other way. And yet, in the nineteenth century, 'non-Euclidean geometry' was invented and eventually applied by Einstein to understand gravity (space is not flat, as Euclid believed, but curved). We might add that, in the 1920s, a certain Louis de Broglie developed a new way of thinking about motion, with laws quite different from those of Newton.

Quantum physics and quantum philosophy

Kant's claims about Newtonian physics are as implausible as his claims about Euclidean geometry. And yet, in the 1920s, Bohr and Heisenberg took Kant's claims about Newtonian physics very seriously. According to them, when we describe a quantum experiment we *have to* use what they called 'the concepts of daily life'—which they identified with classical physics. By 'concepts' they meant something like 'mental framework'. When we set up an experiment in the lab, to describe clearly what we are doing we have to apply the concepts of Newtonian physics (with some updates from Einstein, and from Maxwell's theory of electromagnetism). In 1931 Bohr went so far as to claim that this would remain the case forever:

> The unambiguous interpretation of any measurement must be essentially framed in terms of the classical physical theories, and we may say that in this sense the language of Newton and Maxwell will remain the language of physicists for all time.

Even a reader unfamiliar with these issues may well be puzzled by those last three words. Scientists are committed to reasoning from experimental evidence, and should always bear in mind that our current ideas could turn out to be wrong. So how can a scientist confidently claim that such and such will remain true 'for all time'? The answer lies in Kant's influence: Bohr believed that human nature itself conditions our thinking in ways from which there is no escape.[91]

Heisenberg made similar claims in his book *Physics and Philosophy* (published in 1958):

> The concepts of classical physics are just a refinement of the concepts of daily life and are an essential part of the language which forms the basis of all natural science. . . . There is no use in discussing what could be done if we were other beings than we are. . . . [W]e cannot escape . . . the necessity of using the classical concepts.

In other words, for as long as we remain human, we have no choice but to think about the world in terms of classical physics. For Heisenberg, any

---

[91] Bohr is known to have been deeply influenced by Kant, in part through the writings of the Danish philosopher Harald Høffding.

unambiguous description of an experiment has to make use of classical ideas, without which our science could not continue:

> These concepts are the only tools for an unambiguous communication about events, about the setting up of experiments and about their results. If therefore the atomic physicist is asked to give a description of what really happens in his experiments, the words 'description' and 'really' and 'happens' can only refer to the concepts of daily life or of classical physics. As soon as the physicist gave up this basis he would lose the means of unambiguous communication and could not continue in his science.

This is rather strange. We saw in Chapter 3 that de Broglie was the first person to describe the experiment that we now call electron diffraction or electron interference. And he did not describe it in terms of classical physics but in terms of his new pilot-wave theory. We *can* describe quantum experiments with the concepts provided for us by de Broglie—concepts which differ radically from those of classical physics (as laid down by Newton, Maxwell, and Einstein). So what did Heisenberg mean?

The general reader, who has not suffered a course in quantum physics, may well be puzzled. Heisenberg seems to regard 'concepts of daily life' and 'classical physics' as synonymous. And yet the ancient Greeks and Romans lived daily life for centuries—including the successful construction of bridges, aqueducts, and giant catapults—without understanding classical physics. So, again, what did Heisenberg mean?

It is all very Kantian: clear statements about physical events must employ certain concepts, and the only relevant concepts that human beings can employ are those of classical physics. It is all from Kant—and it is all wrong.

In the Prologue we saw how the Copernican revolution scored a major historical victory for reality over perception, and how quantum mechanics was widely seen as having undone that revolution—putting humanity, with their perceptions, back at the centre of the universe. But historically there was more to this. When Kant's book appeared in 1781 it was widely interpreted—especially in Germany—as having in a sense undone the Copernican revolution. In this new philosophy, the human mind somehow shaped reality and therefore was in a sense at the centre of reality.[92] So the philosophical upheaval that took place in physics in the 1920s had in fact been brewing for a long time. The philosophical reversal of the Copernican revolution, which took place in Germany in the late eighteenth century, was eventually brought into physics by Bohr and Heisenberg. The thinking subject—'the observer'—was now at the centre of physics and of the universe.

We can perhaps see why, in this age-old conflict, the supporters of perception believed they had reached complete and final victory. The laws of physics themselves dictate that we cannot see or pass beyond here. And deep philosophy tells us that our minds cannot even think or conceive beyond here. Never mind lions: beyond here lies madness. So this time there is no hope of escape, no hope of rerunning a Copernican revolution. We have reached the 'inherent limits of knowledge'—with human perception at the centre of the universe. And this time it will stay that way, forever.

Small wonder, then, that to this day there is a strange reluctance among many physicists to question quantum mechanics. Contemporary theoretical physics abounds in boldly speculative theories about high-energy particle physics and the very early universe, including such ideas as granular space, strings vibrating in ten dimensions, endless 'landscapes' of universes with different properties, eternally inflating universes giving birth to more universes, and so on. And yet all of these theories are written down using the laws of quantum mechanics. While there are signs of a recent shift in attitude among some high-energy physicists, the belief is still widespread that somehow quantum mechanics itself is an ultimate theory—even though it makes no ultimate sense—and that to challenge it is not merely difficult or wrong but meaningless.

The two worst theories in history

The last time humanity believed something like this was in the Middle Ages. As we saw in the Prologue, Ptolemaic astronomy put the earth firmly at the centre of the universe—rather like the observer in quantum mechanics. It also divided the universe into a physical sublunary or

---

[92] Whether or not Kant himself intended or believed this is, however, a matter for debate.

earthly realm which we could understand and an abstract heavenly realm which we could not—much as quantum mechanics divides the universe into a physical macroscopic realm which we can understand and an abstract microscopic realm which we cannot. Ptolemaic astronomy worked with uncanny accuracy, in the sense of correctly predicting observable events (such as a solar eclipse). But it provided little in the way of understanding. Much the same can be said of textbook quantum mechanics. Ptolemy's theory was eventually overthrown by Copernicus and Kepler. Some similar revolution surely awaits quantum theory. The question is not if, but when.

Ptolemy lived in the second century CE, in the bustling city of Alexandria, one of the great centres of learning of the ancient world. The Great Library housed the accumulated knowledge and wisdom of the Greeks. The collection may have included a copy of Aristarchus' book—now lost, though cited by Archimedes—proposing a heliocentric theory of the Solar System. But Ptolemy's geocentric theory was to reign supreme for the next 1400 years. It is sobering to recall an epigram famously attributed to Ptolemy, beautifully expressing the sense of transcendence that has driven the true scientific vocation down the ages:[93]

Mortal though I be, yea ephemeral, if but a moment
I gaze up to the night's starry domain of heaven,
Then no longer on earth I stand; I touch the Creator,
And my lively spirit drinketh immortality.

Ptolemy felt he had touched eternal truth. And yet, his theory was catastrophically wrong in almost every detail. To break Ptolemy's spell required the penetrating insight of Copernicus, and the soaring spirit of Kepler, in the sixteenth and seventeenth centuries. By his imagination and intuition, Kepler transformed an abstract geometry of the sky into a material physics of the sky, opening up 'the heavens' to scientific investigation. In a similar vein, modern quantum physicists sometimes voice a sense of touching something deep, perhaps even mystical, in their theoretical system—and yet it all seems egregiously incoherent and wrong. Some feat comparable to that of Copernicus and Kepler will surely be required to break the current deadlock in theoretical physics.

[93] As translated by the English poet Robert Bridges.

We might ask what it was that kept Ptolemy's followers so stuck in their ways. The assumption of a central Earth was certainly one reason. More insidious, perhaps, was the belief that the heavens were intrinsically incomprehensible to mere mortals. And, surely, the killer blow was the related belief that all motion in the sky had to be described in terms of 'perfect' uniform circular motion—making it necessary to introduce the bizarre device of epicycles (Figure 5). These beliefs made it impossible to rethink planetary motion, in terms of elliptical orbits around the Sun, as done much later by Kepler. We might also ask what it is that keeps quantum physicists so stuck in their ways. The assumption of a central observer is certainly one reason. More insidious, perhaps, is the belief that 'the quantum system' is intrinsically incomprehensible to the human mind. And, surely, the killer blow is the related belief that all quantum experiments have to be described in terms of the Kantian concepts of classical physics—making it necessary to accept all sorts of bizarre paradoxes. These beliefs make it impossible to rethink microscopic motion, in terms of trajectories guided by waves, as done long ago by de Broglie.

We are looking at what are probably the two worst widely accepted theories in scientific history—Ptolemaic astronomy and quantum mechanics. In both theories we find abstract mathematical rules which we can apply to make accurate predictions, but which we do not understand in physical terms. There is no clear sense of what is real. And the lack of understanding is excused by a philosophy claiming that certain things are simply beyond human comprehension. Dramatic progress was made in astronomy only when Kepler started to think about the Solar System as a real and concrete physical system, with the Sun exerting a mechanical pull on the planets. To make progress in understanding quantum systems, we must think about them as real and concrete physical things, and find new ways to explain their behaviour. In the 1920s de Broglie did just that. He showed that we can invent new concepts, a new kind of physics, with which we *can* speak about and understand the microworld. In contrast, Bohr and Heisenberg (with Kant in the wings) insisted that our physical thinking could not apply to atoms—much as Ptolemaic astronomers (with Aristotle in the wings) insisted that our physical thinking could not apply to the heavens. For the rest of the twentieth century, de Broglie's theory languished in dark corners of university libraries, among the musty old volumes of the early Solvay conferences, unread and misunderstood even by historians.

The flight from reality and determinism

At this point it seems fair to ask what motivation or satisfaction might be found in setting limits to human understanding in this way. What was in it for them?

In Kant's case there were two main philosophical motivations. The first was concerned with the reliability of human knowledge, while the second attempted to address difficult questions of morality and ethics. It is worth taking a look at these in turn, to understand how and why we have ended up where we are.

We have said that, according to Kant (at least as widely interpreted), Newton's laws tell us less about the structure of physical reality than they do about the structure of human thought. To the uninitiated this may sound bizarre. Surely Newton's physics tells us about the external world and not about how the human mind works. Why would anyone think otherwise? But Kant had his reasons. Kant was famously awakened from what he called a 'dogmatic slumber' by reading the work of the Scottish philosopher David Hume, in particular his book *An Enquiry Concerning Human Understanding* (published in 1748). Hume argued convincingly that we can never be sure that the Sun will rise tomorrow (no matter how many times it has risen in the past), or that when we drop a ball it will fall to the ground (no matter how many times that has happened in the past), and so on. The observed laws of physics may have been valid yesterday and today, but there is no *logical* reason why they must continue to be valid tomorrow. This 'problem of induction' has bedevilled the philosophy of science ever since, as it casts doubt on the reliability of our scientific knowledge. So now Kant found himself caught in a dilemma. On the one hand it seemed to him that Hume had a point and that scientific knowledge could never be certain. On the other hand, like many others of his time, Kant believed that Newton had discovered the final truth about the physical world. This raised a question: how had Newton done it? Hume shows us by cogent logic that scientific knowledge is unreliable, and yet Newton has discovered final scientific truth. How? Kant's answer was most ingenious. Newton had, in effect, read off the final truth not from nature but from the structure of the human mind. Newtonian physics does not describe reality 'in itself' but only what the human mind can think or understand about reality. As Kant put it (in his *Prolegomena to Any Future Metaphysics*, published in 1783):

> . . . the highest legislation for nature must lie in our self . . . we must not seek the universal laws of nature from nature by means of experience . . . *the understanding does not draw its* (a priori) *laws from nature, but prescribes them to it.*[94]

Astonishing as it may seem, according to Kant our most basic laws of physics (and specifically Newton's laws) are not found in nature by means of experiment but are instead imposed on nature by the human self (or thinking human subject). While we can appreciate the difficult philosophical problem faced by Kant after his reading of Hume, from a modern scientific standpoint his proposed solution seems unacceptable—even absurd.

And now to Kant's second motivation. Living in the aftermath of the Newtonian revolution, Kant believed there could be no place for morals and ethics in a deterministic universe governed by strict laws of cause and effect. If our actions are already determined (or fixed) in advance, it seems we can have no moral responsibility for those actions. These were no idle concerns. They still preoccupy us today. For example, if someone is genetically predisposed to violence, are they responsible for a violent crime they have committed? For Kant there could be no true ethics without freedom of choice.

Kant famously wrote that two things filled him with admiration and reverence:

> . . . the starry heavens above me and the moral law within me.

The question was how to reconcile the two. According to Kant there could be no alternative to Newton's laws since these were in effect fixed features of the human mind. Kant himself had claimed, in an earlier book of 1755, that these laws allowed us to understand the formation of our solar system from a primordial cloud of dust coalescing by gravitational attraction. Were human beings not also part of the material universe—'the starry heavens above'—described by Newtonian science? If so how could that be reconciled with the 'moral law within'? Newton's laws tell us that every event in the material universe—whether it be the motion of a planet across the sky, or the movement of your eyes across this page—is already

---

[94] Italics in the original.

determined in advance. If human beings are part and parcel of the material universe, it seems to follow that there is no freedom of choice. For Kant this was an ethical disaster. And so Kant faced another dilemma: on the one hand he believed that Newtonian physics was true and final as an account of the physical world; on the other hand he thought that Newtonian determinism conflicted with human freedom and ethics. And once again Kant was able to provide a most ingenious answer: that while we can and must reason about the wider world in terms of Newton's laws, in a certain sense we cannot and should not understand ourselves in the same way. In effect, the reasoning human subject is not really part of the material universe as understood by Newtonian physics. Kant's argument is elaborate—and difficult to follow. It appears in his *Critique of Practical Reason* (published in 1788) and has inspired a vast commentary and multiple interpretations.

Today Kant's ethical concerns seem fair enough. But his answer—that our physical theories somehow do not apply to ourselves, in particular when we are making decisions—seems implausible. Our scientific understanding of human nature has made huge strides since the late eighteenth century. We now understand ourselves to be products of biological evolution by natural selection. The physical and biochemical structure of the human body, including the brain, is being mapped with exquisite precision. Brain scans reveal the inner workings of human decision making. Subtle chemical imbalances are known to radically alter mood, memory, and perception. There seem to be no discernible limits to the scientific understanding of human behaviour (even if much remains to be understood). The limits set by Kant seem increasingly incredible. And, finally, other philosophical answers can be and have been given to Kant's ethical questions (though not without controversy).

But here we are interested not so much in the extent to which Kant may or may not have been correct, but rather in how Kant's ideas influenced physical science in the early twentieth century. While Kant is mostly unfamiliar to the general public in the English-speaking world, in continental Europe he is widely seen as a towering cultural and philosophical icon. The publication of Kant's 1781 *Critique of Pure Reason* caused a sensation in Germany, notoriously setting off a profound shift towards subjectivism—the belief in the primacy of human perception and subjective experience, as opposed to objective or material reality (Figure 100). The split between English-speaking 'analytical philosophy' and European 'continental philosophy' can be largely traced to the aftermath of this momentous event.

Figure 100. The rise of German idealism. Kant (left), with the publication of his 1781 *Critique of Pure Reason* (title page, centre), inspired Fichte (right) to argue that human consciousness somehow gives rise to the material universe.

In the late eighteenth and early nineteenth centuries, Kant's ideas inspired the rise of German idealism and German Romanticism, which rapidly became dominant forces in philosophy, literature, art, and music (and eventually politics). The power and influence of post-Kantian thinking was felt especially in Germany—as well as in Bohr's native Denmark.

Kant's writings opened up an intellectual space where serious commentators, such as the influential German philosopher Johann Gottlieb Fichte, could think and write as if the human mind and human spirit somehow came *before* the material universe and in a sense gave rise to it—as if our thoughts and ideas were the primal stuff and the material universe only secondary. As the British philosopher Roger Scruton put it:

> According to the post-Kantian vision, adopted in one form or another by Schiller, Schelling, and Fichte, our world is a world of appearances, organised by our mental faculties; we have no knowledge of things in themselves, and the one certain reality is the self and its freedom.

All this was part and parcel of the philosophical background inherited by Bohr and Heisenberg—who brought something of these ideas into the foundations of modern physics. Kant's influence on Bohr and Heisenberg is well documented (as we have seen). Less well known is Heisenberg's appreciation for the radically subjectivist philosophy of Fichte. In works published in the 1790s, Fichte argued that the world we inhabit is in some sense brought into being by the activity of our consciousness (the 'I' that 'posits itself'). As pointed out by the Israeli-American historian of science Mara Beller, something of Fichte's ideas can be found in Heisenberg's writings. In the concluding paragraph of his famous 1927 paper on the uncertainty principle, we find the enigmatic statement:

> . . . everything observed is a selection from a plenitude of possibilities and a limitation on what is possible in the future.[95]

Heisenberg later confided to his close associate, the German physicist Carl Friedrich von Weizsäcker, that he had 'taken this statement from Fichte'. In a lecture on the history of physical science, delivered at the Academy of Sciences of Saxony and published in 1933, Heisenberg claimed that the very development of our scientific understanding was analogous to Fichte's 'self-limitation of the ego', which Heisenberg explained in words almost identical to those in his uncertainty paper:

> . . . in every act of perception we select one of the infinite number of possibilities and thus we also limit the number of possibilities for the future.

We already know that, for Heisenberg, a quantum system acquires a definite state only when observed by a human experimenter. On this point he was quite explicit. In a particularly telling exchange with Dirac at the 1927 Solvay conference, Heisenberg argued that the outcome of a quantum experiment is somehow brought about, or created, by the act of perception. Heisenberg also seems to have believed that something comparable takes place as science develops. The activity of the mind is not discovering reality but creating it. That such ideas were taken seriously, in the Germany of the 1930s, gives us a sense of how far scientific thinking had drifted from the traditional conception of science rooted in experiment—with the aim of discovering the underlying reality of nature.

---

[95] To technical readers this might appear as simply a statement about the application of probability theory in physics. But Heisenberg's word 'selection' suggests that the observer plays an active role and does not merely record random outcomes. (A more literal translation of the original German reads: 'all perceiving is a choice from a plenitude of possibilities'.)

Perhaps we can now begin to discern the plausibility of the strange world conjured up by quantum physicists in the 1920s: the central role of the observer with their sensory perceptions, the observer's need to interpret experiments in terms of classical physics, the unknowability of atoms in themselves, the only certainty being the observer and their freedom to make measurements and thereby 'create reality'. All this seemed plausible in the eyes of those influenced by Kant as well as by the post-Kantian ideas of Fichte and other founders of German idealism.

In his book of 1958 Heisenberg also claimed that, in quantum physics, we had finally overcome the basic distinction between mind and matter—a distinction often associated with the seventeenth-century French philosopher René Descartes. This 'Cartesian dualism' had already been decisively rejected by German idealist philosophers like Fichte. Certainly, in quantum mechanics, the idea of a material reality independent of the human mind seems to have been lost, and we find ourselves struggling even to ask what is 'really happening' inside quantum systems. In our time some regard this as a step forward, and welcome the blurring of the distinction between mind and matter, seeing it as a vindication of ancient Eastern (Indian or Chinese) philosophy.

In 1975 the Austrian high-energy particle theorist Fritjof Capra published his best-selling book *The Tao of Physics*, in which he drew extensive parallels between modern physics and Eastern mysticism. By his own account, as a student Capra had been deeply impressed and inspired by Heisenberg's book of 1958, in particular the claims about physics having overcome Cartesian dualism. Heisenberg in turn told Capra of his 'complete agreement' with the basic thesis of Capra's book. And so, with Heisenberg, we find ourselves in the same mystical terrain as the later Bohm—unable or unwilling to distinguish reality from our thoughts about reality. As Heisenberg put it in his book, quantum mechanics

. . . makes the sharp separation between the world and the I impossible.

For as long as quantum mechanics continues to work—in the sense of agreeing with experiment, despite its conceptual incoherence—there will be those who believe that 'naïve realism' has been transcended, and that the 1920s witnessed the birth of a new scientific era in which it is meaningless to ask what is actually happening without the presence of a human observer.

By their acts of perception, human beings play a central role in creating the world, which would not meaningfully exist without their presence.

What should we believe? Perhaps something in the human spirit is inclined to turn away from an objective and material reality. The Copernican revolution was and remains one of the most profound shocks ever delivered to the human psyche, leaving us adrift in the vast universe with no apparent purpose. Kant's philosophy attempted to undo that revolution by putting the human mind at the centre of the universe. And something of Kant's teaching was brought into atomic physics by Bohr and Heisenberg. The result is a heady concoction of physics and philosophy, in which the basic distinction between perception and reality has become lost.

Perhaps it is fair to say that modern physics has become so deeply confused that we find ourselves no longer able to think clearly about objective reality. This is welcome news to some, a disaster to others. In the early twentieth century, the deepest level of physics was consigned to the realm of the incomprehensible and unknowable, with scientists in effect retreating once more to a sublunary Ptolemaic world of the senses. And that is where we find ourselves today.

Beyond the quantum

When ancient scientists gazed at the night sky, they thought they saw an ethereal realm beyond human comprehension. If we could send them a word of advice from our own time, it would surely be this: hold your nerve. The world around you can be understood, not only in the earthly realm but far beyond. In the early twentieth century, scientists once again lost their nerve. A combination of wrong arguments and dubious philosophy convinced them there was wisdom in surrender. We had reached the ultimate limits of human knowledge and understanding: microscopic atoms and elementary particles inhabit an ethereal and perhaps even mystical realm beyond human comprehension. An awkward timidity took hold. To think and inquire further was to be naïve, to acquiesce was to be sophisticated.

As we survey the walls of our intellectual prison—built for us by Bohr and Heisenberg, with bricks and mortar supplied by Kant and Fichte—we look for some weak spot, some small sign that we might be able to escape. But how?

We need to look long and hard at quantum mechanics. It is cunningly contrived. The theory speaks only of human perceptions—'measurements made by the observer'—with no clear statements about objective reality. It somehow manages to make correct predictions, while maintaining a studied silence about what is actually happening. We are perplexed. Then we notice something. Entangled particles seem to be instantaneously connected across empty space—even though Einstein's relativity teaches us that this is impossible. We are told not to worry: we cannot use entanglement to send an actual signal, so there is no genuine conflict with relativity. Even so, something does not seem quite right. If entangled particles really are instantaneously connected over large distances, why can we not harness that connection to send superluminal signals? It is as if there is some sort of conspiracy: something extraordinary is going on behind the scenes, which we are never allowed to see or control. Why?

Then we remember. In the nineteenth century theorists were worried about the 'heat death of the universe'. One day, in the far future, the stars would all burn out and everything in the universe would reach the same temperature. In that peculiar state of heat death, it would no longer be possible to convert heat into work. Imaginary beings in such a world would be inclined to call this a law of physics. Some of them, however, might see it as a suspicious conspiracy. The closer they looked, the deeper the conspiracy would become. Everything in their world is infused with the same thermal noise or microscopic jittering, making it impossible for them to control the finer details of molecular motion. Why? Eventually some of them might understand that these limitations are not laws of physics after all but mere peculiarities of the state they happen to be living in—and that there is a whole new physics beyond what they know, a physics in which heat can be harnessed as a source of power and the obscuring effects of thermal noise can be overcome.

And so now *we* wonder: might something similar have happened already in our world? Could there be a deeper reality behind the scenes, a reality that is hidden from us simply because we are in a comparable state of 'quantum death'—where everything is infused with the same quantum noise or microscopic jittering, making it impossible for us to control the finer details of particle motion? Perhaps Heisenberg's uncertainty principle is not a law of physics after all but merely a peculiarity of the state we happen to be living in—and there is a whole new physics beyond what we know, a physics of 'quantum life' in which entangled particles can be harnessed for superluminal signalling and the obscuring effects of quantum noise can be overcome.

That sounds like an idea, but how to develop it? We ponder a good while, trying this and that, imagining what might happen if we could escape from quantum death. Certainly we expect to be able to signal faster than light. But without a concrete theory it is difficult to make progress. Then one day we come across a musty old volume in a dark corner of the physics library. It is the proceedings of a conference held in 1927, where Louis de Broglie presented his 'pilot-wave theory', a theory that was revived and extended by David Bohm in 1952. We are astonished to discover that the theory describes exactly what we have in mind—even if de Broglie and Bohm themselves did not quite see it that way.

Pilot-wave theory tells us that quantum systems are real things containing particles (or fields) moving around in definite ways. The motion is guided by a pilot wave (the usual quantum wave). The precise trajectories determine the outcomes of quantum experiments—whether or not an observer is present. There is no need for 'quantum mysticism'. In practical experiments we work with large numbers of similar systems, and we are interested in the overall statistical distribution of outcomes—such as the percentages of particles landing near various points in a two-slit experiment. The results depend on how the particle positions are distributed at the beginning of the experiment. *If* the initial distribution agrees with Born's formula, so that probabilities are given by the squared magnitude of the quantum wave, then the final results are the same as in quantum mechanics. We find the usual distribution of outcomes, and the statistics obey the uncertainty principle. In other words, if we assume the usual quantum noise at the beginning, we obtain the usual quantum noise at the end—and we are then unable to see or control the underlying trajectories. That much is uncontroversial.

Since the 1950s, the small number of workers actively supporting pilot-wave theory have left it there. By taking Born's probability formula as a law or axiom, the results are the same as those of standard quantum mechanics. But that is like insisting that everything in the universe must always be at the same temperature, so that the state of heat death is the only possible state—in which case all systems always have the same thermal noise. The insistence on Born's formula has continued down to our own day, with most workers in the field concluding that in pilot-wave theory there is no observable new physics beyond what is already given in the quantum textbooks.

And so there it is, staring us in the face: the key that can unlock the door of our prison. As we have argued in this book, Born's formula is *not* a law of physics. It merely describes an equilibrium state of quantum death which we happen to be confined to. In that special state, quantum noise and the uncertainty principle reign supreme. But there is a whole new physics beyond what we know, in which quantum noise can be evaded. In this new nonequilibrium physics, Born's formula fails and the uncertainty principle can be beaten. Microscopic reality can be clearly seen and controlled—and entangled particles *can* be deployed for superluminal signalling. Quantum life beckons.

In this book we have argued that, behind the obscure formulas of quantum mechanics, there lies hidden a lost world—which can be described precisely by pilot-wave theory, and in which remote objects are connected directly and instantaneously across the reaches of space. That world is closed to us now because we live in a state of quantum death, in which all the matter surrounding us is pervaded by quantum noise (described by Born's formula)—a statistical fog which was probably created soon after the Big Bang. To access the deeper new physics requires us to peer behind the fog. This is not easy.

Evidence for the new physics might be found in the cosmic microwave background, which contains imprints from the very early universe—possibly from before the fog was formed. Perhaps we will discover relic particles which were created close to the Big Bang. Such particles could still break Born's formula today. Alternatively, at the deepest level of physics, we have argued that gravity may have the power to 'revive' systems from the state of quantum death, thereby creating particles or fields that break Born's formula—for example, in radiation from evaporating primordial black holes, radiation which is being searched for by satellites in space.

The discovery of such particles—a radically new form of matter—would allow us to penetrate the quantum fog and escape from quantum death once and for all. We could harness subquantum matter for seemingly impossible tasks such as superluminal signalling and eavesdropping on quantum secret codes. This would undoubtedly usher in a new technological age, as far-reaching as that initiated by the discovery of electricity and magnetism.

But only future research can tell if any of the theoretical scenarios described in this book will succeed in practice.

In ancient times, some astronomers had suggested that the earth is rotating daily on its axis, instead of the Sun and other stars rotating daily around the earth. In his *Almagest*, Ptolemy was scrupulous enough to consider this idea and to argue against it. According to his reasoning, if the earth was rotating at such a speed, a falling body would appear to move rapidly sideways as it fell, since the ground itself would be moving rapidly in the opposite direction. We observe no such thing and so, Ptolemy concluded, the earth cannot be rotating. In these fascinating passages of Ptolemy's book, we catch a glimpse of what must have been lively disagreements among ancient astronomers, with some suggesting that the air and terrestrial objects were somehow carried around by the rotating earth. It seemed to Ptolemy that a great conspiracy would be required to prevent us from noticing that the earth on which we are standing is moving at great speed. We now know that, in fact, no conspiracy is needed: our modern understanding of mechanics explains why we do not feel the motion of the earth. Similarly, today, many critics of pilot-wave theory object that we are unable to measure the details of the underlying trajectories. If microscopic particles and fields really are moving as de Broglie and Bohm claimed, why do we not notice? It is as if a great conspiracy is required to prevent us from seeing reality as it is. But, again, no conspiracy is needed. If we understand pilot-wave theory correctly, we find that quantum noise arises naturally from a process of quantum relaxation as we have described, and that in the resulting state of quantum death the details of microscopic motion are obscured from view. Pilot-wave theory provides a simple explanation for why we do not see the trajectories, just as modern mechanics provides a simple explanation for why we do not feel the motion of the earth.

What is real? It is a question that lies at the heart of science. It is arguably the most fertile question ever posed by the human mind. The answers found thus far have revealed wonders beyond our wildest imagination, from the molecular structure of life to the vast expanses of intergalactic space. Despite the sceptics down the ages, the question remains as vital as ever.

Glossary

absolute time. The true time valid for all observers.

action at a distance. When an object or particle affects another instantaneously across empty space with no intervening medium or field.

astrophysics. The physics of stars and galaxies.

Bell's inequality. A mathematical restriction on the correlation function for spin measurements made on entangled pairs of particles, assuming locality to be true.

Bell's theorem. Proved by Bell in 1964, the theorem shows that a realistic explanation for quantum statistics requires entangled particles to communicate instantaneously across space.

Big Bang. The gigantic explosion at the beginning of our universe.

black hole. A region of space with matter so dense and gravity so intense that not even light can escape.

black hole evaporation. As first theorised by Hawking, owing to quantum effects a black hole can emit radiation and steadily lose its mass.

Born's formula. A basic law of quantum mechanics, whereby the probability of finding a particle near a certain point is given by the squared magnitude of the quantum wave at that point. For a system of many particles, the point is in configuration space. Mathematically, for a system with a quantum wave Ψ, the probability density is $P = |\Psi|^2$ ('P equals mod-psi-squared', the only formula in this book).

Brownian motion. The incessant jittering of tiny granules suspended in water, caused by molecular impacts.

cascade emission. A two-step atomic decay, emitting a pair of entangled photons.

classical physics. Physics as developed by Galileo, Newton, Maxwell, and Einstein (from the seventeenth century to the early twentieth century). This physics makes clear statements about what the world is made of, and provides strict laws of cause and effect.

collapse of the quantum wave. The mysterious process whereby a superposition (or combination) of multiple definite states collapses to only one definite state.

configuration space. A higher-dimensional space in which a single point specifies the complete configuration of a system (for example, it specifies the positions of all the particles in a multi-particle system).

Copernican astronomy. Theory of the Solar System with planets orbiting the Sun.

correlation function. A way of quantifying the tendency for different variables to be related statistically.

cosmic microwave background (CMB). Relic radiation left over from the Big Bang, with a temperature today of about 3 degrees above absolute zero.

cosmology. The study of the universe as a whole.

curved space. A space where straight lines move closer together (as on the surface of a sphere) or move further apart (as on the surface of a saddle), in contrast with flat space where straight lines can remain at the same distance (parallel lines never meet).

dark energy. A mysterious field or substance thought to drive the acceleration of our expanding universe.

dark matter. A mysterious component of galaxies whose extra gravitational force explains why galaxies remain bound together (among other observations).

de Broglie's law of motion. Specifies how a pilot wave determines the velocity of a particle. Roughly, the particle moves perpendicular to the wave crests with a speed inversely proportional to the wavelength. Technically, the velocity is proportional to the phase gradient.

de Broglie's theory of motion (pilot-wave theory). A theory in which a pilot wave causes velocity, as developed by de Broglie in the 1920s.

decoupling (in cosmology). When particles in the early universe no longer interact (after the expanding universe has cooled sufficiently).

definite state. A state with only one value of some quantity (such as energy). Technically, an 'eigenstate'.

determinism. The principle of cause and effect, whereby the future is determined by initial conditions.

diffraction. The tendency of waves to spread around obstacles, provided their wavelength is not too small.

Einstein's theory of gravity. Explains gravitation as an effect of the curvature of space (or spacetime).

electromagnetic field. A field in space with both electric and magnetic components. The field exerts a force on charged particles.

electromagnetic waves. Ripples in the electromagnetic field, travelling at the speed of light. At long wavelengths they are radio waves, at short wavelengths they are X-rays, and in between they are visible light.

electron. A subatomic particle, carrying a negative electric charge, commonly found in orbit around atomic nuclei.

energy levels. Certain allowed values of energy (for example, in atomic systems).

entangled particles. Particles whose joint quantum wave cannot be decomposed into separate waves one for each particle. This means that the particles cannot be considered as truly separate, even when they are far apart.

equilibrium state. A state that, once reached, is maintained over time.

ether. An older word for the electromagnetic field, often with material connotations and commonly regarded as at absolute rest.

event horizon. A surface from behind which not even light can escape.

exoplanet. A planet beyond our solar system.

force. In classical physics, the cause of acceleration.

galaxy. A conglomeration of billions of stars, gravitationally bound together.

gamma rays. Very-high-energy electromagnetic waves, similar to X-rays but of even shorter wavelength.

geocentric. In astronomy, a theory with the earth at the centre of the universe.

gravitational collapse. When matter becomes so dense that an event horizon forms around it, from which not even light can escape.

gravitational force. In Newtonian physics, a piece of matter here exerts a gravitational pull on a piece of matter over there, with a force varying as the inverse square of the distance.

gravitino. A hypothetical subatomic particle associated with gravitational theories of supersymmetry.

Hamiltonian. Technically, the energy 'operator' appearing in the Schrödinger equation.

heat death (of the universe). Theorised in the nineteenth century. In the far future, when all systems have reached the same temperature, it will no longer be possible to convert heat into useful work.

heliocentric. In astronomy, a theory with the Sun at the centre of the universe.

hidden variables. Invisible quantities, whose details are unknown to us, that determine the outcomes of seemingly random quantum events.

hidden-variables theory. A theory with hidden variables.

high-energy physics. Physics at energies so high that relativity is important, where particles can be created and destroyed, and which is usually described by quantum field theory.

idealism (in philosophy). The theory that the universe is ultimately all mind (as opposed to matter).

indefinite state. A state with more than one value of some quantity (such as energy). Technically, a 'superposition of eigenstates'.

inflation (in cosmology). A rapid exponential expansion of space in the very early universe.

inflaton decay. The early process whereby the energy of the inflaton field is converted into matter and radiation, resulting in a 'hot Big Bang'.

inflaton field. A hypothetical field whose potential energy drives inflation.

information loss. The mysterious process whereby an evaporating black hole appears to forget details of the initial state prior to gravitational collapse.

initial conditions. Physical parameters (such as position, velocity, etc.) whose initial values determine the future.

interference (of waves). When two or more waves overlap and combine.

interference pattern. The pattern produced when different waves overlap.

inverse-square law. In Newton's theory of gravity, the attractive force between two masses varies inversely as the square of the distance (so if the masses are twice as far apart, the force is four times less).

kinetic theory of gases. The theory that gases are made of atoms or molecules in rapid motion, the gas pressure being explained by atomic or molecular impacts on the walls of the container.

light quantum. An older word for 'photon'.

light year. The distance light travels in a year (about six trillion miles).

local action. When an object or particle acts locally, that is, only on neighbouring objects or particles.

locality. When interactions occur only between neighbouring objects or particles.

Lorentz invariance. A mathematical symmetry whereby the laws of physics appear the same for all observers in uniform relative motion, provided times and distances are changed appropriately from one observer to another.

Lorentz transformation. A transformation of times and distances that ensures all observers in uniform relative motion experience the same laws of physics.

materialism (in philosophy). The theory that the universe is ultimately all matter (as opposed to mind).

Maxwell's formula. A basic result of kinetic theory, from which we can calculate the probability of finding molecules with given speeds, inside a gas in thermal equilibrium (at a fixed temperature).

measurement problem. The key ambiguity of quantum physics. The absence of a clear dividing line between the macroscopic world of definite states and the microscopic world of indefinite states.

mechanics. The theory of particles or bodies in motion, usually understood to mean classical mechanics (as laid down by Newton and revised by Einstein).

momentum. A measure of motion, specifically mass times velocity.

Newton's first law. A free body moves at uniform speed in a straight line.

Newton's laws of motion. Three laws that define Newton's theory of mechanics.

Newton's second law. The acceleration of a body is proportional to the force acting on it (and inversely proportional to its mass).

Newton's theory of gravity. The theory that matter acts directly across empty space, exerting an attractive force on distant matter, with a magnitude varying as the inverse square of the distance.

Newton's theory of motion (or Newtonian mechanics). A theory in which forces cause acceleration.

nonequilibrium particles. Particles that violate Born's formula.

nonlocality. Direct instantaneous action across empty space, with no intervening medium or field.

non-Newtonian theory of motion. A theory of motion, or of mechanics, that is not based on Newton's laws (for example, de Broglie's theory).

non-orthogonal states. In quantum mechanics, states with a common component.

peaceful coexistence between relativity and quantum mechanics. The two theories are opposed in principle, but live side by side in practice. In particular, despite its underlying nonlocality, quantum mechanics does not permit superluminal signalling.

photon. Often called a 'particle' of light, but more accurately regarded as an energetic excitation of the electromagnetic field.

pilot wave. A guiding wave that determines the motion of particles (or fields).

pilot-wave theory. De Broglie's theory of motion, in which a pilot wave determines or causes velocity.

polarisation (of light or radiation). For a polarised electromagnetic wave, the electric field oscillates along a preferred direction.

power deficit (in cosmology). An unexpected dip in the power spectrum, widely reported at large spatial scales.

power spectrum (in cosmology). A measure of statistical fluctuations from the average, and how they depend on spatial scale, applied in particular to small non-uniformities (of density, temperature, etc.).

preferred rest frame. A true state of rest, commonly conceived with accompanying measuring instruments including clocks.

pre-quantum state. As theorised in this book, a state of matter or radiation showing violations of Born's formula, having not yet reached quantum death.

primordial black hole. A (small) black hole created by the collapse of density perturbations in the early universe.

primordial perturbations. Small non-uniformities in the very early universe.

primordial relic particles. Freely streaming particles, left over from the very early universe, which still exist today.

primordial spectrum. The power spectrum of primordial perturbations.

problem of time (in quantum gravity). The difficulty of accounting for the apparent passage of time in a theory that is fundamentally timeless.

proton. A positively charged subatomic particle, commonly found inside atomic nuclei.

Ptolemaic astronomy. Theory of the Solar System with the Sun, the Moon, the stars, and the other planets orbiting the earth.

quantum computer. A computer whose basic operations depend on quantum mechanics.

quantum conspiracy. As highlighted in this book, the remarkable fact that the nonlocality of quantum physics cannot be employed for superluminal signalling.

quantum cosmology. A theory of the universe as described by quantum mechanics.

quantum cryptography. A form of secret messaging based on quantum key distribution.

quantum death (of the universe). Our present state (as theorised in this book) in which all particles have the same quantum noise, described by Born's formula, making it impossible to harness entangled particles for superluminal signalling.

quantum electrodynamics. The quantum theory of the electromagnetic field and its interaction with charged particles at high energies.

quantum equilibrium. A state of universal quantum noise in which probabilities for particle positions (or field configurations) are forever given by Born's formula.

quantum field theory. The description of fields by quantum mechanics.

quantum gravity. The description of gravity by quantum mechanics.

quantum instability. A gravitationally induced instability of Born's formula.

quantum interference. Conflicting processes seem to occur simultaneously and affect the observed statistical outcomes (for example, in a two-slit experiment).

quantum internet. A global communication system based on quantum physics.

quantum key distribution. The sharing of secret keys (for use in cryptography) by means of quantum effects.

quantum life. As theorised in this book, a state in which Born's formula is violated, so the usual quantum noise does not apply. Entangled particles can be harnessed for superluminal signalling and the uncertainty principle can be evaded.

quantum mechanics. The most fundamental theory of science, first applied to atoms and molecules, and successfully extended to all known particles, fields, and interactions (except gravity).

quantum noise. Statistical noise associated with quantum equilibrium and described by Born's formula.

quantum nonequilibrium. A state in which probabilities for particle positions (or field configurations) violate Born's formula.

quantum nonlocality. When entangled objects or particles seem to be directly connected instantaneously no matter how far apart they may be.

quantum physics. Physics based on quantum mechanics.

quantum probability. Probability as calculated from Born's formula.

quantum relaxation. The process whereby quantum equilibrium is reached, starting from a prior state of quantum nonequilibrium.

quantum system. Any system or thing that is described by quantum mechanics.

quantum theory. Another name for quantum mechanics.

quantum wave. The wave associated with a physical system (such as a particle, a collection of particles, or a field). It is known to physicists as the 'wave function'. In quantum mechanics it is used to calculate probabilities. In pilot-wave theory it guides the motion of individual particles or fields.

qubit. A quantum bit, a superposition of 1 and 0.

randomness. Often called indeterminism. When events take place purely by chance, with no underlying cause.

realism. The philosophical theory that an actual world exists regardless of whether or not human beings are there to observe it.

realist. An adherent of realism.

reference frame. A system of measuring rods and clocks with respect to which we can describe motion.

relative time. Time for, or relative to, a particular observer.

relativity. A theory in which uniform motion is relative, not absolute. Formulated by Galileo ('Galilean relativity') and revised by Einstein ('special relativity'). To avoid confusion, in this book Einstein's later 'general relativity' is called Einstein's theory of gravity.

relaxation. The process whereby equilibrium is reached, starting from a prior state of nonequilibrium.

relic particles. In cosmology, particles that were created in the early universe and which still exist more or less untouched today.

rest frame. A reference frame considered to be at rest.

scalar perturbations. Small non-uniformities in the curvature of space.

Schrödinger equation. The equation for the quantum wave and how it changes.

secret key. In cryptography, a key or code used to encrypt and decrypt messages.

simultaneity (relative or absolute). The property of events occurring at the same time.

singlet state. A special quantum wave with two entangled particles, for which spin measurements along the same axis always yield opposite results.

singularity. A point or region where physical quantities become infinitely large, signalling a breakdown of known physical laws.

spacetime. A four-dimensional amalgamation of space and time.

spectral line. Radiation emitted or absorbed by an atom or molecule close to a particular frequency.

spin. A quantum property of subatomic particles that is formally analogous to the 'angular momentum' of a classical spinning object.

statistical correlation. Measures the extent to which different random events are associated (in a statistical sense).

statistical isotropy (in cosmology). When statistical properties of objects in the sky do not depend on the direction we observe.

subjectivism. The theory that personal human perceptions are the ultimate reality.

sublunary sphere. In Ptolemaic astronomy, the region below the Moon.

subquantum computer. A hypothetical computer with components built from subquantum matter, enabling novel forms of computation that are impossible for classical or quantum computers.

subquantum matter. A new form of matter (theoretically possible in pilot-wave theory) that violates Born's formula.

subquantum measurement. A measurement, implemented with subquantum matter, that circumvents the uncertainty principle, allowing us to observe quantum particles without collapsing the quantum wave.

subquantum particles. Particles that violate Born's formula.

subquantum physics. New physics beyond quantum mechanics, enabled by violations of Born's formula (theoretically possible in pilot-wave theory).

superluminal. Faster than the speed of light.

supermassive black hole. A black hole of at least a million times the mass of the Sun, commonly found at the centres of galaxies.

superposition. A combination of multiple definite states which in some sense exist simultaneously.

tensor perturbations. Ripples in the curvature of space (gravitational waves).

thermal equilibrium. A state of uniform temperature, in which probabilities for molecular speeds are given by Maxwell's formula.

thermal noise. Statistical noise associated with thermal equilibrium and described by Maxwell's formula.

thermal nonequilibrium. A state of non-uniform temperature, or a state with no well-defined temperature at all, in which probabilities for molecular speeds violate Maxwell's formula.

thermal relaxation. The process whereby thermal equilibrium is reached, starting from a prior state of thermal nonequilibrium.

thermodynamics. The science of heat.

two-slit experiment. An iconic quantum experiment in which, it is widely claimed, a single particle traverses two different holes at the same time.

uncertainty principle. A basic limitation of quantum physics, according to which we cannot measure both the position and the velocity of a particle at the same time.

wave function. The technical term for what this book calls the 'quantum wave'.

Wheeler–DeWitt equation. The equation for the quantum wave when gravity is included.

X-rays. High-energy electromagnetic waves of short wavelength.

Bibliography

Frequently Cited Works

Abraham, Colin, and Valentini (2014): E. Abraham, S. Colin, and A. Valentini, Long-time relaxation in pilot-wave theory, *Journal of Physics A* 47, 395306 (2014).

Bacciagaluppi and Valentini (2009): G. Bacciagaluppi and A. Valentini, *Quantum Theory at the Crossroads: Reconsidering the 1927 Solvay Conference* (Cambridge University Press, Cambridge, 2009).

Bell (1987): J. S. Bell, *Speakable and Unspeakable in Quantum Mechanics* (Cambridge University Press, Cambridge, 1987).

Bohm (1952a): D. Bohm, A suggested interpretation of the quantum theory in terms of 'hidden' variables. I, *Physical Review* 85, 166 (1952).

Bohm (1952b): D. Bohm, A suggested interpretation of the quantum theory in terms of 'hidden' variables. II, *Physical Review* 85, 180 (1952).

Colin and Valentini (2013): S. Colin and A. Valentini, Mechanism for the suppression of quantum noise at large scales on expanding space, *Physical Review D* 88, 103515 (2013).

Colin and Valentini (2015): S. Colin and A. Valentini, Primordial quantum nonequilibrium and large-scale cosmic anomalies, *Physical Review D* 92, 043520 (2015).

de Broglie (1928): L. de Broglie, La nouvelle dynamique des quanta, in: *Électrons et Photons: Rapports et Discussions du Cinquième Conseil de Physique* (Gauthier-Villars, Paris, 1928). [English translation: Bacciagaluppi and Valentini (2009).]

Holland (1993): P. R. Holland, *The Quantum Theory of Motion: An Account of the de Broglie-Bohm Causal Interpretation of Quantum Mechanics* (Cambridge University Press, Cambridge, 1993).

Kandhadai and Valentini (2020): A. Kandhadai and A. Valentini, Mechanism for nonlocal information flow from black holes, *International Journal of Modern Physics A* 35, 2050031 (2020).

Pearle and Valentini (2006): P. Pearle and A. Valentini, Quantum mechanics: generalizations, in: *Encyclopaedia of Mathematical Physics*, eds. J.-P. Françoise et al. (Elsevier, North-Holland, Amsterdam, 2006).

Struyve and Valentini (2009): W. Struyve and A. Valentini, De Broglie-Bohm guidance equations for arbitrary Hamiltonians, *Journal of Physics A* 42, 035301 (2009).

Towler, Russell, and Valentini (2012): M. D. Towler, N. J. Russell, and A. Valentini, Time scales for dynamical relaxation to the Born rule, *Proceedings of the Royal Society A* 468, 990 (2012).

Underwood and Valentini (2015): N. G. Underwood and A. Valentini, Quantum field theory of relic nonequilibrium systems, *Physical Review D* 92, 063531 (2015).

Valentini (1991a): A. Valentini, Signal-locality, uncertainty, and the subquantum H-theorem. I, *Physics Letters A* 156, 5 (1991).

Valentini (1991b): A. Valentini, Signal-locality, uncertainty, and the subquantum H-theorem. II, *Physics Letters A* 158, 1 (1991).

Valentini (1992): A. Valentini, On the pilot-wave theory of classical, quantum and subquantum physics, PhD thesis, International School for Advanced Studies, Trieste, Italy (1992). [http://hdl.handle.net/20.500.11767/4334]

Valentini (1996): A. Valentini, Pilot-wave theory of fields, gravitation and cosmology, in: *Bohmian Mechanics and Quantum Theory: An Appraisal*, eds. J. T. Cushing et al. (Kluwer, Dordrecht, 1996).

Valentini (2001): A. Valentini, Hidden variables, statistical mechanics and the early universe, in: *Chance in Physics: Foundations and Perspectives*, eds. J. Bricmont et al. (Springer, Berlin, 2001).

Valentini (2002a): A. Valentini, Subquantum information and computation, *Pramana–Journal of Physics* 59, 269 (2002).

Valentini (2002b): A. Valentini, Signal-locality in hidden-variables theories, *Physics Letters A* 297, 273 (2002).

Valentini (2004): A. Valentini, Black holes, information loss, and hidden variables, arXiv:hep-th/0407032.

Valentini (2007): A. Valentini, Astrophysical and cosmological tests of quantum theory, *Journal of Physics A* 40, 3285 (2007).

Valentini (2008): A. Valentini, Hidden variables and the large-scale structure of space-time, in: *Einstein, Relativity and Absolute Simultaneity*, eds. W. L. Craig and Q. Smith (Routledge, London, 2008).

Valentini (2009): A. Valentini, Beyond the quantum, *Physics World* 22N11, 32 (2009).

Valentini (2010): A. Valentini, Inflationary cosmology as a probe of primordial quantum mechanics, *Physical Review D* 82, 063513 (2010).

Valentini (2014): A. Valentini, Trans-Planckian fluctuations and the stability of quantum mechanics, arXiv:1409.7467 [hep-th].

Valentini (2020): A. Valentini, Foundations of statistical mechanics and the status of the Born rule in de Broglie-Bohm pilot-wave theory, in: *Statistical Mechanics and Scientific Explanation: Determinism, Indeterminism and Laws of Nature*, ed. V. Allori (World Scientific, Singapore, 2020).

Valentini (2021): A. Valentini, Quantum gravity and quantum probability, arXiv:2104.07966 [gr-qc].

Valentini (2023): A. Valentini, Beyond the Born rule in quantum gravity, *Foundations of Physics* 53, 6 (2023).

Valentini (2024a): A. Valentini, De Broglie-Bohm quantum mechanics, in: *Encyclopedia of Mathematical Physics*, 2nd edn (Elsevier, Amsterdam, 2024).

Valentini (2024b): A. Valentini, Pilot-wave theory and the search for new physics, *Annales de la Fondation Louis de Broglie* 48, 329 (2024).

Valentini (2025): A. Valentini, De Broglie-Bohm pilot-wave theory, in: *Oxford Research Encyclopedia of Physics* (Oxford University Press, Oxford, 2025). [https://oxfordre.com/physics]

Valentini (forthcoming): A. Valentini, *Introduction to Quantum Foundations and Pilot-Wave Theory* (Oxford University Press, Oxford, forthcoming).

Valentini and Westman (2005): A. Valentini and H. Westman, Dynamical origin of quantum probabilities, *Proceedings of the Royal Society A* 461, 253 (2005).

Vitenti, Peter, and Valentini (2019): S. Vitenti, P. Peter and A. Valentini, Modeling the large-scale power deficit with smooth and discontinuous primordial spectra, *Physical Review D* 100, 043506 (2019).

Front Matter

E. A. Poe, Mesmeric Revelation, in: *Collected Works of Edgar Allan Poe, vol. III, Tales and Sketches 1843–1849*, ed. T. O. Mabbott (The Belknap Press, Cambridge, MA, 1978).

Acknowledgements

E. Schrödinger, *Science and Humanism: Physics in Our Time* (Cambridge University Press, Cambridge, 1951).

PROLOGUE

What is real?

It has been said that nobody understands it ('I think I can safely say that nobody understands quantum mechanics'): R. P. Feynman, *The Character of Physical Law* (Modern Library/BBC, London, 1965). [Reprinted in 2017 by MIT Press, p. 129.]

Reality and marginalisation: A. Becker, *What is Real? The Unfinished Quest for the Meaning of Quantum Physics* (Basic Books, New York, 2018).

F. Guicciardini, *Ricordi* (1528), extract as translated in: L. Barzini, *The Italians* (Penguin Books, London, 1968), p. 9.

The eternal conflict: perception vs. reality

Passionate historic debate about kinetic theory: L. S. Feuer, *Einstein and the Generations of Science* (Basic Books, New York, 1974), pp. 19–20, 335–341.

Lenin and kinetic theory: V. I. Lenin, *Materialism and Empirio-criticism* (Wellred Books, London, 2021).

Mach's philosophy of sensations: E. Mach, *The Analysis of Sensations, and the Relation of the Physical to the Psychical* (The Open Court Publishing Company, Chicago and London, 1914). [Reprinted: Dover Publications, New York, 1959]

Boltzmann's philosophy of atomism: L. Boltzmann, *Theoretical Physics and Philosophical Problems*, ed. B. McGuinness (Reidel, Dordrecht, 1974).

Mach's 1910 denial of kinetic theory: E. Mach, The guiding principles of my scientific theory of knowledge and its reception by my contemporaries, English translation in: *Physical Reality: Philosophical Essays on Twentieth-Century Physics*, ed. S. Toulmin (Harper & Row, New York, 1970). The full quotation reads: 'If belief in the reality of atoms is so important to you, I cut myself off from the physicist's mode of thinking, I do not wish to be a true physicist, I renounce all scientific respect—in short: I decline with thanks the communion of the faithful.'

Perception vs. reality in astronomy: the Copernican revolution

Ptolemaic system of astronomy: C. Ptolemy, *Ptolemy's Almagest*, trans. & ed. G. J. Toomer (Princeton University Press, Princeton, 1998).

Copernican system of astronomy: N. Copernicus, *On the Revolutions of Heavenly Spheres*, trans. C. G. Wallis (Prometheus Books, Amherst, New York, 1995).

Aristotle's *On the Heavens*: *The Treatises of Aristotle, On the Heavens, On Generation and Corruption, and On Meteors*, trans. T. Taylor (Robert Wilks, London, 1807). [Reprinted: Elibron Classics, 2006.]

Plato's *Timaeus*: *The Dialogues of Plato*, vol. III, trans. B. Jowett (Oxford University Press, London, 1892).

Saving the phenomena: P. Duhem, *To Save the Phenomena: an Essay on the Idea of Physical Theory from Plato to Galileo* (University of Chicago Press, Chicago and London, 1969); M. J. Crowe, Theories of the World: from Antiquity to the Copernican Revolution (Dover Publications, New York, 2001), chapter 5.

Kepler, *New Astronomy* (1609): J. Kepler, *Astronomia Nova*, trans. W. H. Donahue (Green Lion Press, Santa Fe, 2015).

Kepler: A. Koestler, *The Sleepwalkers: A History of Man's Changing Vision of the Universe* (Hutchinson, London, 1959).

Kepler and motive force: A. Koyré, *The Astronomical Revolution: Copernicus—Kepler—Borelli* (Methuen, London, 1973), pp. 190, 197.

Newton, *Principia* (1687): I. Newton, *The Principia: Mathematical Principles of Natural Philosophy*, trans. I. Bernard Cohen & A. Whitman (University of California Press, Oakland, 1999).

Perception vs. reality in quantum physics: undoing the Copernican revolution

Textbook quantum mechanics: P. A. M. Dirac, *The Principles of Quantum Mechanics* (Oxford University Press, Oxford, 1958); R. Shankar, Principles of Quantum Mechanics (Springer, New York, 1994); D. J. Griffiths, Introduction to Quantum Mechanics (Cambridge University Press, Cambridge, 2017).

Schrödinger's cat: E. Schrödinger, Die gegenwärtige Situation in der Quanten-mechanik, *Die Naturwissenschaften* 23, 807 (1935). [English translation: The present situation in quantum mechanics, in: Quantum Theory and Measurement, eds. J. A. Wheeler and W. H. Zurek (Princeton University Press, Princeton, 1983).]

Why reality won the day

Brownian motion: P. Pearle et al., What Brown saw and you can too, *American Journal of Physics* 78, 1278 (2010).

G. Galilei, *Sidereus Nuncius, or The Sidereal Messenger*, trans. A. Van Helden (University of Chicago Press, Chicago/London, 2015).

Why reality will win again

Observing the observer: E. P. Wigner, Remarks on the mind-body question, in: *The Scientist Speculates*, ed. I. J. Good (Heinemann, London, 1961).

But now is not the time?

Bernoulli and kinetic theory: D. Bernoulli, *Hydrodynamica* (Argentorati, 1738).

Herapath and Waterston: S. G. Brush, John James Waterston and the kinetic theory of gases, *American Scientist* 49, 202 (1961).

Lucretius and Brownian motion: Lucretius, *On the Nature of Things* (Dover Publications, New York, 2004).

Aristarchus' heliocentric model: T. Heath, *Aristarchus of Samos: the Ancient Copernicus* (Clarendon Press, Oxford, 1913).

Louis de Broglie's 1927 pilot-wave theory

Key references on pilot-wave theory: de Broglie (1928), Bohm (1952a, b), Holland (1993), Valentini (2024a, 2025, forthcoming).

What this book is about

Key references on pilot-wave theory as understood in this book: Abraham, Colin, and Valentini (2014); Colin and Valentini (2013, 2015); Kandhadai and Valentini (2020); Pearle and Valentini (2006); Struyve and Valentini (2009); Towler, Russell, and Valentini (2012); Underwood and Valentini (2015); Valentini (1991a,b, 1992, 1996, 2001, 2002a, 2004, 2007, 2008, 2009, 2010, 2014, 2020, 2021, 2023, 2024a, 2025, forthcoming); Valentini and Westman (2005); Vitenti, Peter, and Valentini (2019).

1 BEFORE THE QUANTUM

A theory of quantum reality—and its rejection, twice

Einstein's concern about the 'spooky' character of quantum theory (nonlocality and nonseparability) in his discussions with Bohr: D. Howard, 'Nicht sein kann was nicht sein darf', or the prehistory of EPR, 1909–1935: Einstein's early worries about the quantum mechanics of composite systems, in: *Sixty-two Years of Uncertainty*, ed. A. I. Miller (Plenum Press, New York, 1990).

What really happened at the 1927 Solvay conference: Bacciagaluppi and Valentini (2009). See, in particular, chapter 12 'Beyond the Bohr-Einstein debate'.

De Broglie's pilot-wave theory and its revival by Bohm: de Broglie (1928), Bohm (1952a,b)

Nonlocality: or spooky action at a distance

Nonlocality in pilot-wave theory: Bell (1987), chapter 1; Holland (1993).

A strange conspiracy

Conspiratorial features of pilot-wave theory: Valentini (1991b, 1992, 1996).

Heat death of the universe: or what happens when the lights go out

W. Thomson, On a universal tendency in nature to the dissipation of mechanical energy, *Philosophical Magazine* 4, 304 (1852).

R. Clausius, *The Mechanical Theory of Heat* (John Van Voorst, London, 1867), pp. 290–1, 364–5.

Quantum death of the universe: why reality is invisible

Valentini (1991a,b, 1992, 1996, 2020, forthcoming).

Quantum relaxation: how quantum death happens

Valentini (1992); Valentini and Westman (2005); Towler, Russell, and Valentini (2012); Abraham, Colin, and Valentini (2014).

Three ways out of quantum death

The first loophole, relic cosmological particles: Valentini (2001, 2007); Underwood and Valentini (2015).

The second loophole, long wavelengths in the CMB: Valentini (2007, 2010); Colin and Valentini (2013, 2015); Viteni, Peter, and Valentini (2019).

The third loophole, black holes and quantum gravity: Valentini (2004, 2007, 2021, 2023); Kandhadai and Valentini (2020).

A fourth loophole, a breakdown of de Broglie's law at nodes: Valentini (2014, 2024b, forthcoming); A. Valentini and M. Varma, Towards a test of the Born rule in high-energy collisions, arXiv:2505.07510.

2 SEEING THE IMPOSSIBLE

Quantum pornography
C. Dewdney, private communications, 2021 and 2022.

The 'only mystery' of quantum mechanics
The wavy pattern contains 'the *only* mystery' of quantum mechanics: R. P. Feynman, R. B. Leighton, and M. Sands, *The Feynman Lectures on Physics, vol. III: Quantum Mechanics* (Addison-Wesley, Reading, MA, 1965), chapter 1, p.1.

A physics of the invisible
De Broglie's pilot-wave theory and its revival by Bohm: de Broglie (1928), Bohm (1952a, b).

Dewdney's work on the two-slit experiment: C. Philippidis, C. Dewdney and B. J. Hiley, Quantum interference and the quantum potential, *Nuovo Cimento B* 52, 15 (1979); C. Dewdney, *On the Foundations of Quantum Theory*, PhD thesis, Birkbeck College (1983).

Dewdney's own account including his later work: C. Dewdney, Rekindling of de Broglie-Bohm pilot wave theory in the late twentieth century: a personal account, *Foundations of Physics* 53, 24 (2023).

Two books on pilot-wave theory: D. Bohm and B. J. Hiley, *The Undivided Universe: an Ontological Interpretation of Quantum Theory* (Routledge, London, 1993); Holland (1993).

Bell's theorem: J. S. Bell, On the Einstein Podolsky Rosen paradox, *Physics* 1, 195 (1964); Bell (1987), chapter 2.

Bell's critique of quantum mechanics: Bell (1987).

What about quantum randomness?
The uncertainty principle applies only to quantum death: Valentini (1991b, 2002a, forthcoming).

Born's formula and quantum death
Key references on quantum relaxation to Born's formula: Valentini (1991a, 1992, 2020); Valentini and Westman (2005); Towler, Russell, and Valentini (2012); Abraham, Colin, and Valentini (2014).

Explaining nonlocality

Einstein on nonlocality as 'spooky': *The Born-Einstein Letters: Correspondence between Albert Einstein and Max and Hedwig Born from 1916 to 1955 with commentaries by Max Born*, trans. I. Born (MacMillan, London, 1971), p. 158.

Peaceful coexistence between relativity and quantum mechanics: A. Shimony, Controllable and uncontrollable non-locality, in: *Foundations of Quantum Mechanics in the Light of New Technology*, eds. S. Kamefuchi et al. (Physical Society of Japan, Tokyo, 1984).

3 ORIGINS OF PILOT-WAVE THEORY

Louis de Broglie in 1920s Paris

De Broglie's family background: A. Abragam, Louis Victor Pierre Raymond de Broglie, 15 August 1892–19 March 1987, *Biographical Memoirs of Fellows of the Royal Society* 34, 23 (1988); M. J. Nye, Aristocratic culture and the pursuit of science: the de Broglies in modern France, Isis 88, 397 (1997).

De Broglie's argument against Newton's first law, and his prediction of electron diffraction and interference: L. de Broglie, Quanta de lumière, diffraction et interférences, *Comptes Rendus Hebdomadaires des Séances de l'Académie des Sciences (Paris)* 177, 548 (1923).

De Broglie's PhD thesis: L. de Broglie, Recherches sur la théorie des quanta, PhD thesis, University of Paris (1924); L. de Broglie, Recherches sur la théorie des quanta, Annales de Physique (10), 3, 22 (1925).

Einstein's verdict: A. Einstein, letter to Langevin, 16 December 1924, quoted in: O. Darrigol, Strangeness and soundness in Louis de Broglie's early works, *Physis* 30, 303 (1993) [reprinted with typographical corrections, 1994], p. 355.

Schrödinger wrote to Einstein, quoted in: J. Mehra and H. Rechenberg, *The Historical Development of Quantum Theory*, vol. 5, part 2 (Springer, New York, 1987), p. 412.

Schrödinger found the equation for de Broglie's waves: Bacciagaluppi and Valentini (2009), pp. 52–4.

De Broglie's final formulation of pilot-wave theory: L. de Broglie, La mécanique ondulatoire et la structure atomique de la matière et du rayonnement, *Le Journal de Physique et le Radium* (6), 8, 225 (1927); de Broglie (1928).

De Broglie at the Solvay conference

Standard historical accounts of the 1927 Solvay conference: M. Jammer, *The Conceptual Development of Quantum Mechanics* (McGraw-Hill, New York, 1966); J. Mehra, The Solvay Conferences on Physics: Aspects of the Development of Physics since 1911 (Reidel, Dordrecht and Boston, 1975); J. Mehra and H. Rechenberg, The Historical Development of Quantum Theory, vol. 6, part 1 (Springer, New York, 2000).

De Broglie's theory 'was hardly discussed at all' and 'the only serious reaction came from Pauli': M. Jammer, *The Philosophy of Quantum Mechanics: The Interpretations of Quantum Mechanics in Historical Perspective* (John Wiley and Sons, New York, 1974), pp. 110–11.

Published proceedings of the 1927 Solvay conference (in French): *Électrons et Photons: Rapports et Discussions du Cinquième Conseil de Physique* (Gauthier-Villars, Paris, 1928).

English translation and analysis: Bacciagaluppi and Valentini (2009).

Ironies of history

Feynman concedes to Bohm that pilot-wave theory is a logical possibility: O. Freire, *David Bohm: a Life Dedicated to Understanding the Quantum World* (Springer, New York, 2019), p. 94.

Bohm thanks Feynman for 'interesting and stimulating discussions': D. Bohm, Proof that probability density approaches $|\Psi|^2$ in causal interpretation of the quantum theory, *Physical Review* 89, 458 (1953).

The birth of the measurement problem

Pauli, letter to Bohr, August 1927: W. Pauli, *Wissenschaftlicher Briefwechsel mit Bohr, Einstein, Heisenberg u.a., Teil I: 1919–1929*, eds. A. Hermann, K. v. Meyenn, and V. F. Weisskopf (Springer, Berlin, 1979), pp. 404–405. [English translation: Bacciagaluppi and Valentini (2009), p. 55.]

Einstein's 1927 argument: Bacciagaluppi and Valentini (2009), section 7.1.

Popping into existence: see the related video animation by George Musser, https://spookyactionbook.com/2015/10/03/einsteins-bubble-paradox-video/.

Pearle's theory of random collapse: P. Pearle, Reduction of the state vector by a nonlinear Schrödinger equation, *Physical Review D* 13, 857 (1976); P. Pearle, Combining stochastic dynamical state-vector reduction with spontaneous localization, Physical Review A 39, 2277 (1989); P. Pearle, Introduction to Dynamical Wave Function Collapse (Oxford University Press, Oxford, 2024).

Everett's theory of parallel universes: H. Everett, On the foundations of quantum mechanics, PhD thesis, Princeton University (1957); H. Everett, 'Relative state' formulation of quantum mechanics, Reviews of Modern Physics 29, 454 (1957).

Heisenberg on 'ideological superstructure': W. Heisenberg, *Physics and Philosophy: The Revolution in Modern Science* (Harper & Brothers, New York, 1958), pp. 131–133.

Deutsch on pilot-wave theory: D. Deutsch, Interview, in: *The Ghost in the Atom: a Discussion of the Mysteries of Quantum Physics*, eds. P. C. W. Davies and J. R. Brown (Cambridge University Press, Cambridge, 1986), p. 102.

The trouble with pilot-wave theory

Einstein's alternative pilot-wave theory: D. W. Belousek, Einstein's 1927 unpublished hidden-variable theory: Its background, context and significance, *Studies in History and Philosophy of Modern Physics* 27, 437 (1996); P. R. Holland,

What's wrong with Einstein's 1927 hidden-variable interpretation of quantum mechanics?, Foundations of Physics 35, 177 (2005); Bacciagaluppi and Valentini (2009), section 11.3.

Einstein withdraws his paper: A. Pais, Einstein and the quantum theory, *Reviews of Modern Physics* 51, 863 (1979), p. 901, footnote 83; C. Kirsten and H. J. Treder, Albert Einstein in Berlin 1913–1933 (Akademie-Verlag, Berlin, 1979), p. 135.

Einstein withdraws his commitment to speak: A. Pais, *Subtle is the Lord: The Science and the Life of Albert Einstein* (Oxford University Press, Oxford, 1982), p. 432.

Bohm's version is unstable: S. Colin and A. Valentini, Instability of quantum equilibrium in Bohm's dynamics, *Proceedings of the Royal Society A* 470, 20140288 (2014).

The vexing question of Born's formula

Bohm and quantum relaxation: D. Bohm, Proof that probability density approaches $|\Psi|^2$ in causal interpretation of the quantum theory, *Physical Review* 89, 458 (1953).

Bohm and Vigier and a randomly-fluctuating fluid: D. Bohm and J. P. Vigier, Model of the causal interpretation of quantum theory in terms of a fluid with irregular fluctuations, *Physical Review* 96, 208 (1954).

The author and quantum relaxation: Valentini (1991a, b).

The author's PhD thesis: Valentini (1992).

Forever confined to quantum death, circular argument for Born's formula: D. Dürr, S. Goldstein and N. Zanghì, Quantum equilibrium and the origin of absolute uncertainty, *Journal of Statistical Physics* 67, 843 (1992).

Circular reasoning influential among some philosophers: D. Albert, *After Physics* (Harvard University Press, Cambridge, MA, 2015).

Rebuttal and critique of the Bohmian mechanics school: Valentini (2020).

Quantum mysticism: wholeness and nonlocality

Bohm on the 'universal flux': D. Bohm, *Wholeness and the Implicate Order* (Routledge & Kegan Paul, London, 1980), p. 67.

Bohm's dialogues with Krishnamurti: J. Krishnamurti and D. Bohm, *The Ending of Time* (Harper & Row, San Francisco, 1985).

Dewdney at Birkbeck College: C. Dewdney, private communications, 2021 and 2022.

Dewdney's discovery in a bookshop: F. J. Belinfante, *A Survey of Hidden-Variables Theories* (Pergamon Press, Oxford, 1973).

Bohm's new understanding of pilot-wave theory: D. Bohm, B. J. Hiley, and P. N. Kaloyerou, An ontological basis for the quantum theory, *Physics Reports* 144, 321 (1987).

Bohm and Hasted meet and test Geller: J. Hasted, *The Metal-benders* (Routledge & Kegan Paul, London, 1981), pp. 8–11, 18, 136–40; J. Hasted, D. Bohm, E. Bastin, and B. O'Regan, Experiments on psychokinetic phenomena, in: The

Geller Papers: Scientific Observations on the Paranormal Powers of Uri Geller, ed. C. Panati (Houghton Mifflin, Boston, 1976), pp. 183–96.

Bohm and Hasted reply to criticisms: J. Hasted, D. Bohm, E. Bastin, and B. O'Regan, *Nature* 254, 470 (1975).

Spiritualism and the paranormal at Birkbeck: J. Bourke, Radical physics: science, socialism, and the paranormal at Birkbeck College in the 1970s, *Journal of the British Academy* 7, 25 (2019), 45–6.

Spiritualism and the paranormal in British academia: R. Noakes, *Physics and Psychics: The Occult and the Sciences in Modern Britain* (Cambridge University Press, Cambridge, 2019).

Bernard Carr and the paranormal: B. J. Carr, Worlds apart? Can psychical research bridge the gulf between matter and mind?, *Proceedings of the Society for Psychical Research* 59, 1 (2008); B. J. Carr, Hyperspatial models of matter and mind, in: Beyond Physicalism: Toward Reconciliation of Science and Spirituality, eds. E. F. Kelly, A. Crabtree, and P. Marshall (Rowman & Littlefield, Maryland, 2015).

Brian Josephson and the paranormal: B. D. Josephson and F. Pallikari-Viras, Biological utilization of quantum nonlocality, *Foundations of Physics* 21, 197 (1991).

Stewart and Tait and the ether: B. Stewart and P. G. Tait, *The Unseen Universe: or Physical Speculations on a Future State* (MacMillan, London, 1875). [Reprinted: Palala Press, 2016.]

Oliver Lodge and the ether: O. Lodge, *Ether and Reality: a Series of Discourses on the Many Functions of the Ether of Space* (Hodder and Stoughton, London, 1925), p. 179. [Reprinted: Cambridge University Press, 2012.]

Oliver Lodge and spiritualist séances: O. Lodge, *Raymond: or Life and Death, with Examples of the Evidence for Survival of Memory and Affection after Death* (Methuen & Co., London, 1916).

Descartes' vortex theory: R. Descartes, *Principia Philosophiae* (1644). [R. Descartes, Principles of Philosophy, trans. V. R. Miller and R. P. Miller (Kluwer, Dordrecht, 1991).]

Newton's study of alchemy and his rejection of the mechanical philosophy: R. S. Westfall, *The Life of Isaac Newton* (Cambridge University Press, Cambridge, 1993), chapters 6 and 7.

De Broglie and Bergson: L. S. Feuer, *Einstein and the Generations of Science* (Basic Books, New York, 1974), pp. 206–14.

Action at a distance: fact or fantasy?

There must be a deeper reality behind quantum mechanics: A. Einstein, B. Podolsky, and N. Rosen, Can quantum-mechanical description of physical reality be considered complete? *Physical Review* 47, 777 (1935).

Same argument made earlier by Einstein: L. Hardy, The EPR argument and nonlocality without inequalities for a single photon, in: *Fundamental Problems in Quantum Theory*, eds. D. Greenberger and A. Zeilinger (New York Academy of Sciences, New York, 1995); Bacciagaluppi and Valentini (2009), pp. 175–8.

Bohr's reply to EPR: N. Bohr, Can quantum-mechanical description of physical reality be considered complete? *Physical Review* 48, 696 (1935).

Controversy over the EPR argument: G. Bacciagaluppi and E. Crull, *The Einstein Paradox: The Debate on Nonlocality and Incompleteness in 1935* (Cambridge University Press, Cambridge, 2024).

Bell's theorem: J. S. Bell, On the Einstein Podolsky Rosen paradox, *Physics* 1, 195 (1964); Bell (1987), chapter 2.

Bell had read Bohm's papers: Bell (1987), pp. 11, 160.

Bell's inequality at small angles: N. Herbert, Cryptographic approach to hidden variables, *American Journal of Physics* 43, 315–16 (1975); L. E. Ballentine, Quantum Mechanics: A Modern Development (World Scientific, Singapore, 1998), pp. 586–7.

Experimental violations of Bell's inequality: S. J. Freedman and J. F. Clauser, Experimental test of local hidden-variable theories, *Physical Review Letters* 28, 938 (1972); A. Aspect, J. Dalibard, and G. Roger, Experimental test of Bell's inequalities using time-varying analyzers, Physical Review Letters 49, 1804 (1982).

Clauser recalls being summarily dismissed by Feynman: John Clauser, Nobel Prize in Physics 2022, Official interview, December 2022: https://www.nobelprize. org/prizes/physics/2022/clauser/interview.

Loopholes in Bell's theorem: W. Myrvold, M. Genovese, and A. Shimony, Bell's theorem, in: *The Stanford Encyclopedia of Philosophy*, ed. E. N. Zalta (Fall 2021 edition). [https://plato.stanford.edu/archives/fall2021/entries/bell-theorem]

Theories with back-in-time causation: K. B. Wharton and N. Argaman, Colloquium: Bell's theorem and locally mediated reformulations of quantum mechanics, *Reviews of Modern Physics* 92, 021002 (2020).

Superdeterminism as a vast conspiracy: J. S. Bell, Bertlmann's socks and the nature of reality, *Journal de Physique Colloques* 42, C2–41 (1981); Bell (1987), p. 154; I. Sen and A. Valentini, Superdeterministic hidden-variables models I: non-equilibrium and signalling, Proceedings of the Royal Society A 476, 20200212 (2020); I. Sen and A. Valentini, Superdeterministic hidden-variables models II: conspiracy, Proceedings of the Royal Society A 476, 20200214 (2020).

Recent proponents of superdeterminism: S. Hossenfelder and T. Palmer, Rethinking superdeterminism, *Frontiers in Physics* 8, 139 (2020).

'... there must be a mechanism ...': J. S. Bell, On the Einstein Podolsky Rosen paradox, *Physics* 1, 195 (1964); Bell (1987), p. 20.

4 TRANSCENDING RELATIVITY

The end of peaceful coexistence

Superluminal signalling and revising relativity: Valentini (1991b, 2008).

Relativity in a nutshell

Galileo and the relativity of uniform motion: G. Galilei, *Dialogue Concerning the Two Chief World Systems: Ptolemaic and Copernican*, trans. S. Drake (Modern Library, New York, 2001).

One of Lorentz's key papers on relativity: H. A. Lorentz, Electromagnetic phenomena in a system moving with any velocity less than that of light, *Proceedings of the Academy of Sciences of Amsterdam* 6, 809 (1904). [Reprinted in: The Principle of Relativity: A Collection of Original Memoirs on the Special and General Theory of Relativity, trans. W. Perrett and G. B. Jeffery (Dover, New York, 1952).]

Einstein's first paper on relativity: A. Einstein, Zur Elektrodynamik bewegter Körper, *Annalen der Physik* 322, 891 (1905). [English translation, On the electrodynamics of moving bodies, in: The Principle of Relativity: A Collection of Original Memoirs on the Special and General Theory of Relativity, trans. W. Perrett and G. B. Jeffery (Dover, New York, 1952).]

Back to Lorentz: or reversing relativity

Synchronising clocks with instantaneous signals: Valentini (2008).

Lorentz's interpretation of his equations: H. A. Lorentz, Das Relativitätsprinzip und seine Anwendung auf einige besondere physikalische Erscheinungen, in: *Das Relativitätsprinzip: Eine Sammlung von Abhandlungen* (Teubner, Leipzig, 1913).

Poincaré's work on relativity: H. Poincaré, Sur la dynamique de l'électron, *Comptes Rendus de l'Académie des Sciences de Paris* 140, 1504 (1905); H. Poincaré, Sur la dynamique de l'électron, Rendiconti del Circolo Matematico di Palermo 21, 129 (1906); W. Pauli, Theory of Relativity (Pergamon Press, London, 1958), pp. 3, 21; Valentini (2008).

Lorentzian interpretation of time dilation and length contraction: J. S. Bell, How to teach special relativity, in: Bell (1987), p. 67; H. R. Brown, *Physical Relativity: Space-Time Structure from a Dynamical Perspective* (Oxford University Press, Oxford, 2005).

Relativity and pilot-wave theory

Preferred rest frame in pilot-wave theory: A. Valentini, On Galilean and Lorentz invariance in pilot-wave dynamics, *Physics Lett*ers A 228, 215 (1997).

Relativity and quantum death

Dirac's quantum theory of the electromagnetic field: P. A. M. Dirac, The quantum theory of the emission and absorption of radiation, *Proceedings of the Royal Society of London A* 114, 243 (1927).

Pilot-wave theory of the electromagnetic field: Bohm (1952b), appendix A.

Lorentz invariance is broken outside of quantum death: Valentini (1992, forthcoming).

Lorentz invariance in modern quantum field theory: M. E. Peskin and D. V. Schroeder, *An Introduction to Quantum Field Theory* (Addison-Wesley, Reading, Mass., 1995); S. Weinberg, The Quantum Theory of Fields. I. Foundations (Cambridge University Press, Cambridge, 1995).

Possible breakdown of Lorentz invariance at high energies: V. A. Kostelecký and M. Mewes, Signals for Lorentz violation in electrodynamics, *Physical Review D* 66, 056005 (2002); P. Hořava, Quantum gravity at a Lifshitz point, Physical Review D 79, 084008 (2009).

5 BEYOND QUANTUM PHYSICS

The quantum conspiracy—the game is up

Explaining the quantum conspiracy: Valentini (1991b, 1992, 1996, 2002b, forthcoming).

Superluminal signalling

Flipping a quantum coin from far away: D. A. Rice, A geometric approach to non-locality in the Bohm model of quantum mechanics, *American Journal of Physics* 65, 144 (1997).

Superluminal signalling in pilot-wave theory: Valentini (1991b, 2002a, forthcoming).

Superluminal signalling in general hidden-variables theories: Valentini (2002b); A. Valentini, Signal-locality and subquantum information in deterministic hidden-variables theories, in: *Non-locality and Modality*, eds. T. Placek and J. Butterfield (Kluwer, Dordrecht, 2002).

Quantum relaxation: or how the universe got shaken

Quantum relaxation to Born's formula: Valentini (1991a, 1992); Valentini and Westman (2005); S. Colin, Relaxation to quantum equilibrium for Dirac fermions in the de Broglie-Bohm pilot-wave theory, *Proceedings of the Royal Society A* 468, 1116 (2012); Towler, Russell, and Valentini (2012); Abraham, Colin, and Valentini (2014); C. Efthymiopoulos, G. Contopoulos, and A. C. Tzemos, Chaos in de Broglie-Bohm quantum mechanics and the dynamics of quantum relaxation, Annales de la Fondation Louis de Broglie 42, 133 (2017); N. G. Underwood, Extreme quantum nonequilibrium, nodes, vorticity, drift and relaxation retarding states, Journal of Physics A 51, 055301 (2018); A. Drezet, Justifying Born's rule $P=|\Psi|^2$ using deterministic chaos, decoherence, and the de Broglie-Bohm quantum theory, Entropy 23, 1371 (2021); F. B. Lustosa, N. Pinto-Neto, and A. Valentini, Evolution of quantum non-equilibrium for coupled harmonic oscillators, Proceedings of the Royal Society A 479, 20220411 (2023).

Beating the uncertainty principle

Violating Born's formula breaks the uncertainty principle: Valentini (1991b, forthcoming).

Subquantum measurements: Valentini (2002a, forthcoming).

6 A MESSAGE FROM THE BEGINNING OF TIME

Looking backwards

The most distant galaxies found so far: B. E. Robertson et al., Discovery and properties of the earliest galaxies with confirmed distances, arXiv:2212.04480 [astro-ph.GA].

The Big Bang and the cosmic microwave background (CMB): P. Peter and J.-P. Uzan, *Primordial Cosmology* (Oxford University Press, Oxford, 2009).

Discovery of the CMB: A. A. Penzias and R. W. Wilson, A measurement of excess antenna temperature at 4080 Mc/s, *Astrophysical Journal* 142, 419 (1965).

Quantum mechanics in the sky

Guth's seminal paper on inflation: A. H. Guth, Inflationary universe: a possible solution to the horizon and flatness problems, *Physical Review D* 23, 347 (1981).

Quantum wrinkles (or fluctuations) during inflation: S. W. Hawking, The development of irregularities in a single bubble inflationary universe, *Physics Letters B* 115, 295 (1982); A. H. Guth and S.-Y. Pi, Fluctuations in the new inflationary universe, Physical Review Letters 49, 1110 (1982).

Inflationary cosmology: A. R. Liddle and D. H. Lyth, *Cosmological Inflation and Large-Scale Structure* (Cambridge University Press, Cambridge, 2000); V. Mukhanov, Physical Foundations of Cosmology (Cambridge University Press, Cambridge, 2005).

COBE satellite results: G. F. Smoot et al., Structure in the *COBE* differential microwave radiometer first-year maps, *Astrophysical Journal* 396, L1 (1992).

Testing quantum mechanics at the beginning of time: Valentini (2010).

'Our most puzzling finding'

Three-year results from the *WMAP* satellite: D. N. Spergel et al., Three-year Wilkinson Microwave Anisotropy Probe (WMAP) observations: implications for cosmology, *Astrophysical Journal Supplement Series* 170, 377 (2007).

Planck satellite 2013 results, 'our most puzzling finding': Planck Collaboration: P. A. R. Ade et al., *Planck* 2013 results. XV. CMB power spectra and likelihood, *Astronomy and Astrophysics* 571, A15 (2014).

Evidence for a violation of statistical isotropy at large scales: Planck Collaboration: P. A. R. Ade et al., *Planck* 2013 results. XXIII. Isotropy and statistics of the CMB, *Astronomy and Astrophysics* 571, A23 (2014).

The 'axis of evil': K. Land and J. Magueijo, Examination of evidence for a preferred axis in the cosmic radiation anisotropy, *Physical Review Letters* 95, 071301 (2005).

Persistence of statistical anisotropy in the *Planck* data: Planck Collaboration: P. A. R. Ade et al., *Planck* 2015 results. XVI. Isotropy and statistics of the CMB, *Astronomy and Astrophysics* 594, A16 (2016).

Cosmic imprints of quantum relaxation

Quantum relaxation for a field: Valentini (2007).

Quantum relaxation suppressed by the expansion of space at long wavelengths, implying a large-scale power deficit in the CMB: A. Valentini (2008), De Broglie-Bohm prediction of quantum violations for cosmological super-Hubble modes, arXiv:0804.4656 [hep-th]; Valentini (2010); Colin and Valentini (2013).

Nonequilibrium breaks statistical isotropy: Colin and Valentini (2015); A. Valentini (2015), Statistical anisotropy and cosmological quantum relaxation, arXiv:1510.02523 [astro-ph.CO].

A failure of quantum mechanics?

Noise deficit as a function of wavelength: Colin and Valentini (2015); S. Colin and A. Valentini, Robust predictions for the large-scale cosmological power deficit from primordial quantum nonequilibrium, *International Journal of Modern Physics D* 25, 1650068 (2016).

Search for large-scale suppression of quantum noise in CMB data: Vitenti, Peter, and Valentini (2019).

Robert Rines and the Loch Ness Monster: *New York Times* obituary (2009), https://www.nytimes.com/2009/11/08/us/08rines.html.

The mystery of polarisation

Our trawl through the CMB data: Vitenti, Peter, and Valentini (2019).

Breaking the inflationary consistency relation: Valentini (2010).

Two independent violations of Born's formula in the early universe: Vitenti, Peter, and Valentini (2019).

New analysis of CMB polarisation data: Planck Collaboration: Y. Akrami et al., *Planck* 2018 results. VII. Isotropy and statistics of the CMB, *Astronomy and Astrophysics* 641, A7 (2020).

Last word to the *Planck* team: Planck Collaboration: Planck finds no new evidence for cosmic anomalies (6 June 2019), https://sci.esa.int/s/A1G5YnW.

7 RELICS FROM THE EARLY UNIVERSE

Quantum archaeology

Pre-quantum relics from the early universe: Valentini (2001, 2007).

The mystery of dark matter

Opinions vary about dark energy: E. Bianchi, C. Rovelli, and R. Kolb, Is dark energy really a mystery?, *Nature* 466, 321 (2010).

Evidence for dark matter in galaxies: J. Binney and S. Tremaine, *Galactic Dynamics* (Princeton University Press, Princeton, 2008).

Alternative theories of gravity with no dark matter: M. Milgrom, A modification of the Newtonian dynamics as a possible alternative to the hidden mass hypothesis, *Astrophysical Journal* 270, 365 (1983).

Formation of primordial black holes: S. Hawking, Gravitationally collapsed objects of very low mass, *Monthly Notices of the Royal Astronomical Society* 152, 75 (1971).

Dark matter and primordial black holes: B. Carr, F. Kuhnel and M. Sandstad, Primordial black holes as dark matter, *Physical Review D* 94, 083504 (2016); B.

Carr et al., Constraints on primordial black holes, Reports on Progress in Physics 84, 116902 (2021).

Dark matter and relic particles: G. Bertone, D. Hooper and J. Silk, Particle dark matter: evidence, candidates and constraints, Physics Reports 405, 279 (2005); Particle Dark Matter: Observations, Models and Searches, ed. G. Bertone (Cambridge University Press, Cambridge, 2010).

Survivors of quantum death?

Inflaton decay: P. Peter and J.-P. Uzan, *Primordial Cosmology* (Oxford University Press, Oxford, 2009).

Decaying inflaton field can create nonequilibrium particles: Underwood and Valentini (2015).

Supersymmetry and dark matter: J. Ellis and K. A. Olive, Supersymmetric dark matter candidates, in: *Particle Dark Matter: Observations, Models and Searches*, ed. G. Bertone (Cambridge University Press, Cambridge, 2010).

Supersymmetry remains viable: H. Baer et al., Status of weak scale supersymmetry after LHC Run 2 and ton-scale noble liquid WIMP searches, *European Physical Journal Special Topics* 229, 3085 (2020).

Disintegrating dark matter

Disintegrating dark matter: G. Bertone, D. Hooper and J. Silk, Particle dark matter: evidence, candidates and constraints, *Physics Reports* 405, 279 (2005); Particle Dark Matter: Observations, Models and Searches, ed. G. Bertone (Cambridge University Press, Cambridge, 2010).

All particles as energetic excitations of fields: Valentini (1992, 1996, 2024a).

Dirac sea pilot-wave theory of fermions: D. Bohm, B. J. Hiley, and P. N. Kaloyerou, An ontological basis for the quantum theory, *Physics Reports* 144, 321 (1987); D. Bohm and B. J. Hiley, The Undivided Universe: an Ontological Interpretation of Quantum Theory (Routledge, London, 1993); S. Colin, A deterministic Bell model, Physics Letters A 317, 349 (2003); S. Colin and W. Struyve, A Dirac sea pilot-wave model for quantum field theory, Journal of Physics A 40, 7309 (2007); Valentini (2024a).

Dirac's theory of antimatter: P. A. M. Dirac, A theory of electrons and protons, *Proceedings of the Royal Society of London A* 126, 360 (1930).

'Excess' of gamma rays from the centre of our galaxy: The *Fermi* LAT Collaboration, The *Fermi* Galactic Center GeV excess and implications for dark matter, *Astrophysical Journal* 840, 43 (2017).

Recent estimates of the mass of the disintegrating particle: S. Murgia, The *Fermi*-LAT Galactic Center excess: evidence of annihilating dark matter?, *Annual Review of Nuclear and Particle Science* 70, 455 (2020); M. Di Mauro and M. W. Winkler, Multimessenger constraints on the dark matter interpretation of the Fermi-LAT Galactic Center excess, Physical Review D 103, 123005 (2021).

Anomalies in polarisation statistics: A. Valentini, Universal signature of non-quantum systems, *Physics Letters A* 332, 187 (2004).

Testing Born's formula in space

QUICK[3] satellite mission to test Born's formula in space: N. Ahmadi et al., QUICK[3]—Design of a satellite-based quantum light source for quantum communication and extended physical theory tests in space, *Advanced Quantum Technologies* 7, 2300343 (2024).

Distorted spectral lines from violations of Born's formula: N. G. Underwood and A. Valentini, Anomalous spectral lines and relic quantum nonequilibrium, *Physical Review D* 101, 043004 (2020).

Exploding black holes: S. Hawking, Black hole explosions?, *Nature* 248, 30 (1974).

8 SAVED BY GRAVITY

Trapped forever in a quantum fog?

A fourth loophole, a breakdown of de Broglie's law at nodes: Valentini (2014, 2024b, forthcoming); A. Valentini and M. Varma, Towards a test of the Born rule in high-energy collisions, arXiv:2505.07510.

The mystery of gravity and quantum mechanics

Einstein's theory of gravity: A. Einstein, Die Grundlage der allgemeinen Relativitätstheorie, *Annalen der Physik* 354, 769 (1916) [English translation, The foundation of the general theory of relativity, in: The Principle of Relativity: A Collection of Original Memoirs on the Special and General Theory of Relativity, trans. W. Perrett and G. B. Jeffery (Dover, New York, 1952), p. 111]; C. W. Misner, K. S. Thorne, and J. A. Wheeler, Gravitation (W. H. Freeman, San Francisco, 1973).

Quantum electrodynamics: P. A. M. Dirac, The quantum theory of the emission and absorption of radiation, *Proceedings of the Royal Society of London A* 114, 243 (1927); S. S. Schweber, Feynman and the visualization of space-time processes, Reviews of Modern Physics 58, 449 (1986).

Standard model of particle physics: S. Raby, *Introduction to the Standard Model and Beyond* (Cambridge University Press, Cambridge, 2021).

Quantum gravity and the disappearance of time

Quantum gravity and the Wheeler-DeWitt equation: B. S. DeWitt, Quantum theory of gravity. I. The canonical theory, *Physical Review* 160, 1113 (1967); C. Rovelli, Quantum Gravity (Cambridge University Press, Cambridge, 2004); C. Kiefer, Quantum Gravity (Oxford University Press, Oxford, 2012).

Problems with defining a well-behaved time in quantum gravity: W. G. Unruh and R. M. Wald, Time and the interpretation of canonical quantum gravity, *Physical Review D* 40, 2598 (1989); C. J. Isham, Conceptual and geometrical problems in quantum gravity, in: Recent Aspects of Quantum Fields, eds. H. Mitter and H. Gausterer (Springer-Verlag, Berlin, 1991); K. V. Kuchař, Time and interpretations of quantum gravity, in: Proceedings of the 4th Canadian Conference on General Relativity and Relativistic Astrophysics, eds. G. Kunstatter, D. Vincent, and J. Williams (World Scientific, Singapore, 1992) [Reprinted: K. V. Kuchař, International Journal of Modern Physics D 20, 3 (2011)]; C. J. Isham, Canonical quantum gravity and the problem of time, in: Integrable Systems, Quantum Groups, and Quantum Field Theories, eds. L. A. Ibort and M. A. Rodriguez (Kluwer, London, 1993); K. V. Kuchař, The problem of time in quantum geometrodynamics, in: The Arguments of Time, ed. J. Butterfield (Oxford University Press, Oxford, 1999).

The need to reinstate time in quantum gravity: Valentini (1992, 1996); L. Smolin, *Time Reborn: From the Crisis in Physics to the Future of the Universe* (Penguin, 2013).

Physics without time: C. Rovelli, Quantum mechanics without time: a model, *Physical Review D* 42, 2638 (1990); C. Rovelli, Time in quantum gravity: an hypothesis, Physical Review D 43, 442 (1991); C. Rovelli, Forget time, FQXi Essay on the Nature of Time (2009); J. B. Barbour, The timelessness of quantum gravity. I. The evidence from the classical theory, Classical and Quantum Gravity 11, 2853 (1994); J. B. Barbour, The timelessness of quantum gravity. II. The appearance of dynamics in static configurations, Classical and Quantum Gravity 11, 2875 (1994).

Rovelli's later views: C. Rovelli, *The Order of Time* (Penguin, London, 2019).

Hawking's use of Born's formula in quantum gravity: J. B. Hartle and S. W. Hawking, Wave function of the universe, *Physical Review D* 28, 2960 (1983); S. W. Hawking, The quantum state of the universe, Nuclear Physics B 239, 257 (1984); S. W. Hawking and D. Page, Operator ordering and the flatness of the universe, Nuclear Physics B 264, 185 (1986); S. W. Hawking and D. Page, How probable is inflation?, Nuclear Physics B 298, 789 (1988).

Hawking and the reversal of time, mathematically flawed: S. Hawking, *A Brief History of Time: From the Big Bang to Black Holes* (Bantam Press, London, 1988), p. 167.

The naïve interpretation of quantum gravity: W. G. Unruh and R. M. Wald, Time and the interpretation of canonical quantum gravity, *Physical Review D* 40, 2598 (1989).

The problem of time: E. Anderson, *The Problem of Time: Quantum Mechanics versus General Relativity* (Springer, New York, 2017).

Pilot-wave theory to the rescue?

The basic equations of pilot-wave quantum gravity: J. C. Vink, Quantum potential interpretation of the wave function of the universe, *Nuclear Physics B* 369, 707 (1992); T. Horiguchi, Quantum potential interpretation of the Wheeler-DeWitt equation, Modern Physics Letters A 9, 1429 (1994); Yu. V. Shtanov, Pilot wave quantum cosmology, Physical Review D 54, 2564 (1996).

Applications to quantum cosmology by Pinto-Neto and collaborators: N. Pinto-Neto, The Bohm interpretation of quantum cosmology, *Foundations of Physics*

35, 577 (2005); N. Pinto-Neto and J. C. Fabris, Quantum cosmology from the de Broglie-Bohm perspective, Classical and Quantum Gravity 30, 143001 (2013); N. Pinto-Neto, The de Broglie-Bohm quantum theory and its application to quantum cosmology, Universe 7, 134 (2021).

The ekpyrotic universe: J. Khoury, B. A. Ovrut, P. J. Steinhardt, and N. Turok, Ekpyrotic universe: colliding branes and the origin of the hot big bang, *Physical Review D* 64, 123522 (2001).

The problem of probability

Infinite probability and the Wheeler-DeWitt equation: K. V. Kuchař, Time and interpretations of quantum gravity, in: *Proceedings of the 4th Canadian Conference on General Relativity and Relativistic Astrophysics*, eds. G. Kunstatter, D. Vincent and J. Williams (World Scientific, Singapore, 1992) [Reprinted: K. V. Kuchař, International Journal of Modern Physics D 20, 3 (2011)]; K. V. Kuchař, The problem of time in quantum geometrodynamics, in: The Arguments of Time, ed. J. Butterfield (Oxford University Press, Oxford, 1999).

Quantum death transcended

Born's formula is always wrong: Valentini (2021, 2023).

Computer simulations illustrate the absence of quantum relaxation in quantum gravity: A. Kandhadai and A. Valentini, paper in preparation.

How the universe dies

Quantum relaxation in the semiclassical approximation: Valentini (2021, 2023).

The curious case of impossible probability

Corrections to the Schrödinger equation from quantum gravity: C. Kiefer and T. P. Singh, Quantum gravitational corrections to the functional Schrödinger equation, *Physical Review D* 44, 1067 (1991).

Kiefer's book on quantum gravity: C. Kiefer, *Quantum Gravity* (Oxford University Press, Oxford, 2012).

Later derivations of similar impossible terms: C. Kiefer and M. Krämer, Quantum gravitational contributions to the cosmic microwave background anisotropy spectrum, *Physical Review Letters* 108, 021301 (2012); D. Bini, G. Esposito, C. Kiefer, M. Krämer, and F. Pessina, On the modification of the cosmic microwave background anisotropy spectrum from canonical quantum gravity, Physical Review D 87, 104008 (2013); D. Brizuela, C. Kiefer, and M. Krämer, Quantum-gravitational effects on gauge-invariant scalar and tensor perturbations during inflation: the de Sitter case, Physical Review D 93, 104035 (2016); D. Brizuela, C. Kiefer, and M. Krämer, Quantum-gravitational effects on gauge-invariant scalar and tensor perturbations during inflation: the slow-roll approximation, Physical Review D 94, 123527 (2016).

Alternative derivations of the semiclassical approximation: C. Kiefer and D. Wichmann, Semiclassical approximation of the Wheeler-DeWitt equation: arbitrary orders and the question of unitarity, *General Relativity and Gravitation* 50, 66 (2018); L. Chataignier, Gauge fixing and the semiclassical interpretation of quantum cosmology, Zeitschrift für Naturforschung A 74, 1069 (2019); L. Chataignier and M. Krämer, Unitarity of quantum-gravitational corrections to primordial fluctuations in the Born-Oppenheimer approach, Physical Review D 103, 066005 (2021).

Quantum disintegration: or final escape from quantum death

Making sense of the impossible terms in pilot-wave theory: Valentini (2021, 2023).

Theoretical calculations of quantum disintegration, including for evaporating black holes: Valentini (2021, 2023).

Searching for gamma-rays from exploding primordial black holes: M. Ackermann et al, Search for gamma-ray emission from local primordial black holes with the *Fermi* Large Area Telescope, *Astrophysical Journal* 857, 49 (2018).

9 BLACK HOLES AND THE EDGE OF PHYSICS

Black hole explosions

Formation of a black hole: C. W. Misner, K. S. Thorne and J. A. Wheeler, *Gravitation* (W. H. Freeman, San Francisco, 1973).

Preferred slicing of spacetime associated with an absolute time: Valentini (2004, 2008).

Hawking radiation from an evaporating black hole: S. W. Hawking, Particle creation by black holes, *Communications in Mathematical Physics* 43, 199 (1975).

The information crisis

Hawking's landmark paper on information loss: S. W. Hawking, Breakdown of predictability in gravitational collapse, *Physical Review D* 14, 2460 (1976).

Controversy over the puzzle of information loss: G. Belot, J. Earman, and L. Ruetsche, The Hawking information loss paradox: the anatomy of a controversy, *British Journal for the Philosophy of Science* 50, 189 (1999); D. Wallace, Why black hole information loss is paradoxical, in: Beyond Spacetime: The Foundations of Quantum Gravity, eds. N. Huggett, K. Matsubara, and C. Wüthrich (Cambridge University Press, Cambridge, 2020).

String theory: E. Kiritsis, *String Theory in a Nutshell* (Princeton University Press, Princeton, 2019).

Controversial scientific status of string theory: L. Smolin, *The Trouble with Physics: The Rise of String Theory, the Fall of a Science and What Comes Next* (Houghton Mifflin Company, Boston, 2006).

AdS/CFT correspondence: H. Năstase, *Introduction to the AdS/CFT Correspondence* (Cambridge University Press, Cambridge, 2015).

AdS/CFT and information loss: D. Harlow, Jerusalem lectures on black holes and quantum information, *Reviews of Modern Physics* 88, 015002 (2016); A. Almheiri et al., The entropy of Hawking radiation, Reviews of Modern Physics 93, 035002 (2021); S. Raju, Lessons from the information paradox, Physics Reports 943, 1 (2022).

Nonlocal physics of black holes: S. B. Giddings, Black holes in the quantum universe, *Philosophical Transactions of the Royal Society A* 377, 20190029 (2019).

A new approach to information loss

Pilot-wave theory and information loss: Valentini (2004, 2007, 2021, 2023); Kandhadai and Valentini (2020).

Entangled ingoing and outgoing fields: N. D. Birrell and P. C. W. Davies, *Quantum Fields in Curved Space* (Cambridge University Press, Cambridge, 1982).

Testing quantum physics with black holes

Testing Born's formula for particles entangled across a black-hole event horizon: Valentini (2004, 2007).

Time parameter associated with a preferred slicing of spacetime: Valentini (2008).

An entangled state can channel anomalous noise from inside a black hole: Kandhadai and Valentini (2020).

A quantum experiment with a black hole

Shape of iron line probes curvature of space near the surface of a black hole: A. C. Fabian, M. J. Rees, L. Stella and N. E. White, X-ray fluorescence from the inner disc in Cygnus X-1, *Monthly Notices of the Royal Astronomical Society* 238, 729 (1989); C. S. Reynolds and M. A. Nowak, Fluorescent iron lines as a probe of astrophysical black hole systems, Physics Reports 377, 389 (2003).

First observation of broadened and distorted iron line: Y. Tanaka et al., Gravitationally redshifted emission implying an accretion disk and massive black hole in the active galaxy MCG–6–30–15, *Nature* 375, 659 (1995).

Deduction that the black hole is spinning rapidly: L. W. Brenneman and C. S. Reynolds, Constraining black hole spin via X-ray spectroscopy, *Astrophysical Journal* 652, 1028 (2006).

Two-step cascade emission of pairs of entangled photons near a black hole: Valentini (2004, 2007).

Iron line for a black hole in our galaxy: J. M. Miller et al., Evidence of black hole spin in GX 339–4: *XMM-Newton*/EPIC-pn and *RXTE* spectroscopy of the very high state, *Astrophysical Journal* 606, L131 (2004).

Reprocessing the quantum universe

Gravitational collapse as the greatest crisis in physics of all time: C. W. Misner, K. S. Thorne and J. A. Wheeler, *Gravitation* (W. H. Freeman, San Francisco, 1973), chapter 44.

Natural selection on a cosmological scale: L. Smolin, *The Life of the Cosmos* (Oxford University Press, Oxford, 1997).

Black holes reprocess the quantum universe: Valentini (2004, 2021, 2023).

10 BEYOND QUANTUM TECHNOLOGY

The quantum space race

China launches *Micius* quantum satellite: E. Gibney, Chinese satellite is one giant step for the quantum internet, *Nature* 535, 478 (2016).

Google announces quantum supremacy: F. Arute et al., Quantum supremacy using a programmable superconducting processor, *Nature* 574, 505 (2019).

Quantum supremacy has indeed been achieved: A. Morvan et al., Phase transitions in random circuit sampling, *Nature* 634, 328 (2024).

Estimated global market value of quantum computer technology by 2035: https://www.consultancy.uk/news/24361/quantum-computing-market-to-reach-1-trillion-by-2035.

Micius shares pairs of entangled photons between ground stations: J. Yin et al., Satellite-based entanglement distribution over 1200 kilometers, *Science* 356, 1140 (2017).

Micius shares secret keys: J. Yin et al., Entanglement-based secure quantum cryptography over 1,120 kilometres, *Nature* 582, 501 (2020).

Secure quantum data link between Jinan and Qingdao: Jiu-Peng Chen et al, Twin-field quantum key distribution over a 511 km optical fibre linking two distant metropolitan areas, *Nature Photonics* 15, 570 (2021).

Betting on a theory that makes no sense

Physics is fundamentally about information: C. A. Fuchs, Quantum mechanics as quantum information (and only a little more), arXiv:quant-ph/0205039; C. A. Fuchs, Coming of Age with Quantum Information: Notes on a Paulian Idea (Cambridge University Press, Cambridge, 2011); P. Ball, Beyond Weird: Why Everything You Thought You Knew About Quantum Physics is . . . Different (The Bodley Head, London, 2018).

Bell's pithy critique of information: J. S. Bell, Against 'measurement', in: *Sixty-Two Years of Uncertainty: Historical, Philosophical, and Physical Inquiries into the Foundations of Quantum Mechanics*, ed. A. I. Miller (Plenum Press, New York, 1990). [Reprinted: J. S. Bell, Against 'measurement', Physics World 3(8), 33 (1990), p. 34.]

Hacking the quantum internet

Ekert's 1991 protocol: A. K. Ekert, Quantum cryptography based on Bell's theorem, *Physical Review Letters* 67, 661 (1991).

Hacking into Ekert's protocol: Valentini (2002a).

Bennett's 1992 protocol: C. H. Bennett, Quantum cryptography using any two nonorthogonal states, *Physical Review Letters* 68, 3121 (1992).

1984 protocol by Bennett and Brassard: C. H. Bennett and G. Brassard, Quantum cryptography: public key distribution and coin tossing, in: *International Conference on Computers, Systems and Signal Processing* (IEEE, New York, 1984). [Reprinted: Theoretical Computer Science 560(1), 7 (2014).]

Impossible to distinguish non-orthogonal states reliably: M. A. Nielsen and I. L. Chuang, *Quantum Computation and Quantum Information* (Cambridge University Press, Cambridge, 2000), p. 87.

Reliably distinguishing non-orthogonal states by subquantum measurement: Valentini (2002a).

Subquantum computers

Pioneering work of Feynman and Deutsch: R. P. Feynman, Simulating physics with computers, *International Journal of Theoretical Physics* 21, 467 (1982); D. Deutsch, Quantum theory, the Church-Turing principle and the universal quantum computer, Proceedings of the Royal Society A 400, 97 (1985).

Shor's algorithm: P. W. Shor, Algorithms for quantum computation: discrete logarithms and factoring, in: *Proceedings 35th Annual Symposium on Foundations of Computer Science* (IEEE, New York, 1994).

Power of quantum computers as evidence for many worlds: D. Deutsch, *The Fabric of Reality* (Penguin, London, 1997), p. 217.

Entanglement as the cause of the speed-up: R. Jozsa, Entanglement and quantum computation, in: *The Geometric Universe: Science, Geometry, and the Work of Roger Penrose* (Oxford University Press, Oxford, 1998).

Contextuality as the true driving force: M. Howard et al., Contextuality supplies the 'magic' for quantum computation, *Nature* 510, 351 (2014).

Quantum computing in pilot-wave theory: P. Roser, Quantum computation from a de Broglie-Bohm perspective, arXiv:1205.2563 [quant-ph].

Subquantum computers: Valentini (1992), pp. 86–92.

Subquantum computer with a component that distinguishes non-orthogonal states: Valentini (2002a).

Nonlinear resolution of non-orthogonal states implemented on a quantum computer: D. S. Abrams and S. Lloyd, Nonlinear quantum mechanics implies polynomial-time solution for NP-complete and #P problems, *Physical Review Letters* 81, 3992 (1998).

Possible nonlinear terms in Schrödinger's equation: S. Weinberg, Precision tests of quantum mechanics, *Physical Review Letters* 62, 485 (1989).

Is anyone out there?

Most stars have orbiting planets: A. Cassan et al., One or more bound planets per Milky Way star from microlensing observations, *Nature* 481, 167 (2012).

Proxima Centauri b: G. Anglada-Escudé et al., A terrestrial planet candidate in a temperate orbit around Proxima Centauri, *Nature* 536, 437 (2016); A. Witze, Earth-sized planet around nearby star is astronomy dream come true, Nature 536, 381 (2016).

Planets orbiting the star HR 8799: C. Marois et al., Direct imaging of multiple planets orbiting the star HR 8799, *Science* 322, 1348 (2008).

Amino acids in the Murchison meteorite: K. Kvenvolden et al., Evidence for extraterrestrial amino-acids and hydrocarbons in the Murchison meteorite, *Nature* 228, 923 (1970).

Formaldehyde in interstellar space: L. E. Snyder et al., Microwave detection of interstellar formaldehyde, *Physical Review Letters* 22, 679 (1969).

Complex organic molecules in the Galaxy and beyond: M. Guélin and J. Cernicharo, Organic molecules in interstellar space: latest advances, *Frontiers in Astronomy and Space Sciences* 9, 787567 (2022).

Nucleobases found in meteorites: Y. Oba et al., Identifying the wide diversity of extraterrestrial purine and pyrimidine nucleobases in carbonaceous meteorites, *Nature Communications* 13, 2008 (2022).

Laboratory simulations of interstellar space: Y. Oba et al., Nucleobase synthesis in interstellar ices, *Nature Communications* 10, 4413 (2019).

Organic molecules erupting from Enceladus: F. Postberg et al., Macromolecular organic compounds from the depths of Enceladus, *Nature* 558, 564 (2018).

Extremophile organisms on Earth: N. Merino et al., Living at the extremes: extremophiles and the limits of life in a planetary context, *Frontiers in Microbiology* 10, 780 (2019).

Evidence for early development of simple life forms: E. A. Bell et al., Potentially biogenic carbon preserved in a 4.1 billion-year-old zircon, *Proceedings of the National Academy of Sciences of the United States of America* 112, 14518 (2015); D. Papineau et al, Metabolically diverse primordial microbial communities in Earth's oldest seafloor-hydrothermal jasper, Science Advances 8(15) (2022).

Advanced life forms probably exceedingly rare: P. D. Ward and D. Brownlee, *Rare Earth: Why Complex Life is Uncommon in the Universe* (Copernicus Books, New York, 2000).

Multicellular animal life in extreme environments: G. Borgonie et al., Nematoda from the terrestrial deep subsurface of South Africa, *Nature* 474, 79 (2011).

The search for extraterrestrial intelligence

Our radio technology permits interstellar communication: G. Cocconi and P. Morrison, Searching for interstellar communications, *Nature* 184, 844 (1959).

Voyager space probes: T. Folger, Record-breaking Voyager spacecraft begin to power down, Scientific American, 1 July 2022.

Search for Extraterrestrial Intelligence: SETI at 50, Nature 461, 316 (2009).

Breakthrough Listen programme: https://breakthroughinitiatives.org/initiative/1

Listening to nearby galaxies: C. Choza et al., The Breakthrough Listen search for intelligent life: technosignature search of 97 nearby galaxies, *Astronomical Journal* 167, 10 (2024).

Colonise the Galaxy in one or two million years: M. H. Hart, Explanation for the absence of extraterrestrials on Earth, *Quarterly Journal of the Royal Astronomical Society* 16, 128 (1975).

The Fermi paradox: S. Webb, *If the Universe is Teeming with Aliens . . . Where is Everybody?: Seventy-Five Solutions to the Fermi Paradox and the Problem of Extraterrestrial Life* (Springer, New York, 2015).

The question of alien technology

Searches for extraterrestrial laser signals: N. K. Tellis and G. W. Marcy, A search for laser emission with megawatt thresholds from 5600 FGKM stars, *The Astronomical Journal* 153, 251 (2017); D. Lipman et al., The Breakthrough Listen search for intelligent life: searching Boyajian's star for laser line emission, Publications of the Astronomical Society of the Pacific 131, 034202 (2019).

Aliens might communicate by quantum properties of radiation: A. Berera, Quantum coherence to interstellar distances, *Physical Review D* 102, 063005 (2020); M. Hippke, Searching for interstellar quantum communications, Astronomical Journal 162, 1 (2021); A. Berera and J. Calderón-Figueroa, Viability of quantum communication across interstellar distances, Physical Review D 105, 123033 (2022).

Communication across the universe

Entangled pairs can be created at high-energy accelerators: A. J. Barr, Testing Bell inequalities in Higgs boson decays, *Physics Letters B* 825, 136866 (2022).

Entangled pairs can be created by the expansion of space: E. Martín-Martínez and N. Menicucci, Cosmological quantum entanglement, *Classical and Quantum Gravity* 29, 224003 (2012); L. H. Ford, Cosmological particle production: a review, Reports on Progress in Physics 84, 116901 (2021).

Fidelity of quantum communications can be disrupted by gravitational disturbances: D. E. Bruschi, T. C. Ralph, I. Fuentes, T. Jennewein, and M. Razavi, Spacetime effects on satellite-based quantum communications, *Physical Review D* 90, 045041 (2014).

The most recent calculations: A. Berera and J. Calderón-Figueroa, Viability of quantum communication across interstellar distances, *Physical Review D* 105, 123033 (2022).

Earlier calculations by Berera and collaborators: A. Berera, S. Brahma, R. Brandenberger, J. Calderón-Figueroa, and A. Heavens, Quantum coherence of photons to cosmological distances, *Physical Review D* 104, 063519 (2021).

EPILOGUE

The edge of the world

The end-of-the-road thesis: K. Popper, *Quantum Theory and the Schism in Physics* (Unwin Hyman Ltd, 1982), 'Author's Note', p. xvii.

The ultimate limits of human knowledge

Newtonian physics as a map of the structure of the human mind: I. Kant, *Critique of Pure Reason*, trans. W. S. Pluhar (Hackett Publishing Company, Indianapolis, 1996).

Quantum physics and quantum philosophy

Bohr claims we have to employ the concepts of Newtonian physics: N. Bohr, Maxwell and modern theoretical physics, *Nature* 128, 691 (1931). [Reprinted in: Niels Bohr: Collected Works, vol. 6, ed. J. Kalckar (North-Holland, Amsterdam, 1985), p. 357.]

Bohr deeply influenced by Kant: D. Murdoch, *Niels Bohr's Philosophy of Physics* (Cambridge University Press, Cambridge, 1987); J. Faye, Niels Bohr: His Heritage and Legacy: An Anti-Realist View of Quantum Mechanics (Kluwer, Dordrecht, 1991); H. Pringe, Critique of the Quantum Power of Judgment: A Transcendental Foundation of Quantum Objectivity, dissertation, University of Dortmund (Walter de Gruyter, Berlin, 2007).

Heisenberg made similar claims: W. Heisenberg, *Physics and Philosophy: The Revolution in Modern Science* (Harper & Brothers, New York, 1958), pp. 56, 144.

The two worst theories in history

Epigram attributed to Ptolemy: *The Oxford Book of Greek Verse in Translation*, ed. T. F. Higham and C. M. Bowra (Oxford University Press, 1938), Ptolemaeus, 'Starry Heavens without' (No. 621).

The flight from reality and determinism

Hume argued convincingly: D. Hume, *An Enquiry Concerning Human Understanding* (Oxford University Press, Oxford, 2000).

Newtonian physics does not describe reality in itself: I. Kant, *Prolegomena to Any Future Metaphysics*, trans. ed. G. Hatfield (Cambridge University Press, Cambridge, 2004), '§36. How is nature itself possible?', pp. 71–72.

Two things filled him with admiration and reverence: I. Kant, *Critique of Practical Reason*, trans. M. Gregor (Cambridge University Press, 2015), Conclusion, p. 129.

According to the post-Kantian vision: R. Scruton, *The Ring of Truth: The Wisdom of Wagner's Ring of the Nibelung* (Penguin Books, London, 2017), p. 19.

Fichte and the activity of our consciousness: J. G. Fichte, *The System of Ethics: According to the Principles of the Wissenschaftslehre*, trans. ed. D. Breazeale and G. Zöller (Cambridge University Press, Cambridge, 2005).

Fichte's ideas in Heisenberg's writings: M. Beller, *Quantum Dialogue: the Making of a Revolution* (The University of Chicago Press, Chicago, 1999), p. 67.

Concluding paragraph of the 1927 uncertainty paper: W. Heisenberg, Über den anschaulichen Inhalt der quantentheoretischen Kinematik und Mechanik, *Zeitschrift für Physik* 43, 172 (1927). [English translation: The physical content of quantum kinematics and mechanics, in: Quantum Theory and Measurement, eds. J. A. Wheeler and W. H. Zurek (Princeton University Press, Princeton, 1983), p. 83.]

A more literal translation: Bacciagaluppi and Valentini (2009), p. 169.

Heisenberg had 'taken this statement from Fichte': C. F. von Weizsäcker and C. C. Boone, 'Werner Heisenberg', *CrossCurrents*, Vol. 27, No. 4 (Winter 1977–8), p. 419.

Heisenberg and Fichte's 'self-limitation of the ego': W. Heisenberg, *Philosophic Problems of Nuclear Science* (Pantheon Books, New York, 1952), chapter 2, p. 30.

Heisenberg and the act of perception: Bacciagaluppi and Valentini (2009), pp. 169, 449.

Quantum physics overcomes 'the Cartesian partition': W. Heisenberg, *Physics and Philosophy: The Revolution in Modern Science* (Harper & Brothers, New York, 1958), pp. 77–83.

Extensive parallels between modern physics and Eastern mysticism: F. Capra, *The Tao of Physics* (Wildwood House, London, 1975).

Capra and Heisenberg: F. Capra, *Uncommon Wisdom: Conversations with Remarkable People* (Simon and Schuster, New York, 1988), chapter 1.

As Heisenberg put it in his book: W. Heisenberg, *Physics and Philosophy: The Revolution in Modern Science* (Harper & Brothers, New York, 1958), p. 81.

Beyond the quantum

Ptolemy argues against the rotation of the earth: C. Ptolemy, *Ptolemy's Almagest*, trans. & ed. G. J. Toomer (Princeton University Press, Princeton, 1998), p. 45.

Figure Acknowledgements

Figure 1: Image adapted courtesy of Shutterstock.com.

Figure 2: Ernst Mach, 1910, Österreichische Nationalbibliothek, by Charles Scolik, courtesy of Wikimedia Commons. Ludwig Boltzmann, 1902, unknown author, courtesy of Wikimedia Commons.

Figure 3: Ptolemaic system, archives of Pearson Scott Foresman, courtesy of Wikimedia Commons. Copernican system, courtesy of Eric Blackman, University of Rochester (http://www.pas.rochester.edu/~blackman/ast104/copernican9.html).

Figure 4: O. Fine, *L'Esphere du Monde* (Paris, 1549), courtesy of Picryl.com.

Figure 7 : Representation of Ptolemy by André Thevet, *Les Vrais Pourtraits et Vies des Hommes Illustres Grecz, Latins, et Payens* (Paris, 1584), courtesy of Bibliothèque Nationale de France. Anonymous portrait of Copernicus, courtesy of the District Museum, Toruń. Anonymous portrait of Kepler, courtesy of Smithsonian Libraries and Archives. Portrait of Newton, by Sir Godfrey Kneller, 1689, courtesy of the Estate of the Earl of Portsmouth.

Figure 8: Portrait of Bohr, 1922, courtesy of the Nobel Foundation, Stockholm. Portrait of Heisenberg, 1933, courtesy of German Federal Archives. Portrait of Einstein, courtesy of AIP Emilio Segrè Visual Archives. Portrait of Schrödinger, 1933, courtesy of the Nobel Foundation, Stockholm.

Figures 9, 13, 14, 20, 33, 36, 38, 50, 60, 61, 69, 70, 72, 79, 80, 87, 89, 90, 95, 97, and 98: Some images courtesy of Wpclipart.com.

Figure 10: Adapted from an original image by Doug Hatfield, courtesy of Wikimedia Commons.

Figure 12: Sketches of the surface of the Moon, by Galileo Galilei, *Sidereus Nuncius* (Venice, 1610). Portrait of Galileo, by Justus Sustermans, circa 1637, courtesy of the Uffizi Gallery, Florence.

Figure 15: Portrait of de Broglie, 1929, courtesy of the Nobel Foundation, Stockholm. Portrait of Bohm, 1949, Library of Congress, New York World—Telegram and Sun Collection, courtesy of AIP Emilio Segrè Visual Archives.

Figure 17: The 1927 Solvay conference, group portrait, photograph by Benjamin Couprie, Institut International de Physique Solvay, courtesy of AIP Emilio Segrè Visual Archives.

Figure 32: Images courtesy of Chris Dewdney.

Figure 43: Adapted from Bacciagaluppi and Valentini (2009).

Figure 58: Adapted from Valentini and Westman (2005).

Figures 62 and 63: Courtesy of the European Space Agency (2013).

Figures 64 and 65: Adapted from Colin and Valentini (2013).

Figure 67: Courtesy of NASA, ESA, and the Hubble Heritage Team (STScI/AURA).

Figure 68: Courtesy of NASA, DOE, and the Fermi LAT Collaboration.

Figures 70 and 73: Some images adapted courtesy of Freepik.com.

Figure 71: Image adapted courtesy of Imgur.com.

Figure 73: Some images adapted courtesy of Colourbox.com.

Figure 95: Image of Mars, adapted courtesy of NASA, ESA, and the Hubble Heritage Team STScI/AURA.

Figure 99: Artist's illustration of exoplanet OGLE-2005-BLG-390Lb, adapted courtesy of NASA, ESA, and G. Bacon (STScI).

Figure 100: Portrait of Kant, circa 1790, artist unknown, possibly Elisabeth von Stägemann (school of Anton Graff), courtesy of Wikimedia Commons. Title page of 1781 edition of Immanuel Kant's *Critique of Pure Reason*, courtesy of Wikimedia Commons. Portrait of Johann Gottlieb Fichte, engraved by Johann Friedrich Jügel, after a painting by Heinrich Anton Dähling (1808), courtesy of Wikimedia Commons.

Index